Advances of FOOD TECHNOLOGY

(Volume - 1)

Editors

Dr. Farhan Mohiuddin Bhat

Shruti Dubey

Dr. Okram Abemsana Devi

Neha Naijo Areekal

Integrated Publications ™

New Delhi

Published By: *Integrated Publications* TM

Integrated Publications
H. No. - 3 Pocket - H34, Sector - 3,
Rohini, Delhi-110085, India
Email – info@integratedpublications.in

Editors: *Dr. Farhan Mohiuddin Bhat, Shruti Dubey, Dr. Okram Abemsana Devi and Neha Naijo Areekal*

Edition: *1st*

Publication Year: *2023*

Pages: *260*

ISBN: *978-93-5834-006-8*

Book DOI: *https://doi.org/10.22271/int.book.282*

Price: *₹945/-*

Preface

The main purpose of this book is to provide insight about the different technologies associated with post-harvest processing and value addition of agricultural produces. The content of this book focuses on enhancing quality and shelf life of food products by processing technique, as maintaining of food quality is the basic necessity of life, right from the infant to the old age. Food processing attributes have, therefore, direct bearing in the sustenance of life. The scope of Food Technology shall always be on the rise, in order to meet the requirement of nutritious, healthful, delicious and safe food for each and every human being.

Food Processing sector is regarded as sun rise sector with very good job prospects, so this book has been drafted to revolutionize the food habits of the people by offering processed, cooked and securely packaged foods in attractive forms to consumers. Consumption of processed foods is noticeable across almost all sections of the population. As a result, the market for processed food is growing very fast. technologies associated with post-harvest processing and value addition of commonly used cereals, pulses and oil seeds are elaborated in detail. Methods of preservation in perishable fruits and vegetables in different processed forms has been discussed along with examples in this book.

This book has been written keeping in view the broad audience covering a wide range and provide deep insight on the most current development in the field of food science and technology. The concept and Technologies highlighted in this book has an important bearing in reduction of food wastage and improving nutrition by bringing variety in human dietary patterns.

Farhan M Bhat

Acknowledgement

The contents of the courses were finalized after discussion with editor of the book and approval by the publication agency. I thank all of the editor, faculty and researchers who made whole-hearted efforts in preparing/revising the course contents of this book. I place on record special thanks to Dr. Kunal Narwal, PhD Agronomy for his constant encouragement and support. Thanks are due to him and hi the members of his office for assistance and drafting the proposal of this particular book. Special thanks to everyone in publishing team. Hope this book will certainly help faculty in teaching the various courses of Food Science and Technology program.

Contents

S. No	Chapters	Page No.
1.	**Commercial Processing Technology of Different Fruits and Vegetables Preparation of Juices/Purees, Preparation of RTS/Squashes/Syrups** *(Anjali Kumari Jha)*	01-20
2.	**Preparation of Jam/Jelly/Preserve/Candy, Preparation of Toffee/Bar/Cheese** *(Vaishnavi, Mamta, Nivedhika Rai Vaid and Pooja Bhatt)*	21-40
3.	**Preservation by Sugar- Importance- Principles of Gel Formation- Preparation of Jam, Jelly, Marmalades, Preserve, Candy, Glazed and Crystallized Fruits** *(Shruti Mallikarjun Kolur, Ashwini S. and Santosh M Kolur)*	41-53
4.	**Unfermented fruit Beverages: Preparation and Preservation, Squash, RTS Beverages, Nectar, Cordial, Syrup, Crush, Concentrate. Methods of Extraction, Clarification and Pasteurization** *(Raju C.A, Ashoka S, Ganavi B.R and Pavuluri Yasaswini)*	55-74
5.	**Value Addition of Fruit and Vegetables** *(Sanjay Kumar, Ankit Gavri, Pooja Pahal and Renu)*	75-89
6.	**Post-Harvest Handling Operations; Cleaning: Cleaning of Grains, Washing of Fruits and Vegetables, types of Cleaners, Screens, types of Screens, Rotary Screens, Vibrating Screens, Machinery for Cleaning of Fruits and Vegetables (Air cleaners, Washers), Cleaning Efficiency, Care and Maintenance** *(Aayush Singla, Ankit Gavri, Rajvinder Singh and Tanu Chaudhary)*	91-109
7.	**Sorting and Grading: Sorting, Grading, Methods of Grading; Grading-Size Grading, Color Grading, Specific Gravity Grading; Screening, Equipment for Grading of Fruits and Vegetables, Grading Efficiency, Care and Maintenance** *(Pooja Bhatt, Nivedhika Rai Vaid, Vaishnavi and Mamta)*	111-123

8. **Parboiling Process, Changes During Parboiling, Parboiling Methods, Advantages and Disadvantages of Parboiling with Respect to Milling, Nutrition, and Cooking Quality of Grain. Milling: Milling, Polishing, Grinding, Milling Equipment, Dehuskers, Polishers, (Abrasion, Friction, and Water Jet), Flour Milling Machines, Pulse Milling Machines, Grinders, Cutting Machines, Oil Expellers, Machine Efficiency, and Power Requirement** 125-145
(Neha Naijo Areekal, Ramachandran. R, M. Keerthana and Hemamalini)

9. **Bread - Ingredients - Additives - Chemical Leavening Agent - Different types of Bread -Methods of Bread Preparation - Bread Spoilage and Remedies - Quality Aspects of Bread and its Standards** 147-165
(Abinaya S, Abirami K and Guruju Nookaratnam)

10. **Development of New Cereal-Based Products and Development of Cereal Product Quality** 167-183
(Riddhi Rane and Rohit Palghadmal)

11. **Preservation of Meat; Chilling, Freezing, use of Heat, Dehydration, Smoking, Modified Atmosphere Storage, Ionizing Radiation** 185-194
(Disha Chavan, Pravin Zine and Nikhil Kadam)

12. **The Proportion of Muscular tissue in Mutton and Poultry, Comminuted meat Products, Fresh Sausages, Cured Sausages, and Fermented Sausages** 195-212
(Neha Naijo Areekal, S. Menaka and Yogeshwari. R)

13. **Cleaning and Sanitation; Cleaning Procedures, Cleaning agents, Disinfection Techniques, Disinfectants for the Meat Industry, Cleaning and Disinfection Plans** 213-224
(Nivedhika Rai Vaid, Pooja Bhatt, Mamta and Vaishnavi)

14. **Meat Processing Techniques** 225-245
(Akash Khorde, Dhaneshwari Karnavat and Deepti Chaudhari)

15. **Methods of Cooking; Frying, Roasting, Broiling, Braising and Stewing** 247-260
(Mamta, Vaishnavi, Pooja Bhatt and Nivedhika Rai Vaid)

Chapter - 1

Commercial Processing Technology of Different Fruits and Vegetables Preparation of Juices/Purees, Preparation of RTS/Squashes/Syrups

Author

Anjali Kumari Jha

M.Sc Scholar, Department of Horticulture (Vegetable & Floriculture) Bihar Agricultural College, Bihar Agricultural University, Sabour, Bhagalpur, Bihar, India

Chapter - 1

Commercial Processing Technology of Different Fruits and Vegetables Preparation of Juices/Purees, Preparation of RTS/Squashes/Syrups

Anjali Kumari Jha

Abstract

India, though ranks second in the world in production of fruits and vegetables, yet most of its population is deprived of its daily requirement of fruits and vegetables. The estimated losses in fruits and vegetables are higher and reached from 30 to 40 per cent. And only 2% of total production of fruits and vegetables are processed in our country. These percentages are not acceptable and adversely affect the Indian economy. A variety of strategies are suggested to reduce the losses of fruits and vegetables. Processing of fruits and vegetables has been practiced since a long time. This serves a number of purposes such as making the produce available in off season, reducing transportation and storage costs, enhancing shelf life, value addition and of course reduction in postharvest losses. A range of initiatives are made to reduce these losses through the use of appropriate post-harvest management measures. Cleaning, washing, disinfecting, drying, packaging, and storing are all examples. There are several methods of preservation of foods and processing of fruits and vegetables for value addition and to increase shelf life. Thus, processing is of great importance in the fruit and vegetable sector, as it not only increases marketing opportunities, but also adds value and reduces post-harvest losses.

Keywords: *Processing, technology, losses, economy, value addition, shelf life*

Introduction

The variable climate of India ensures the availability of a wide range of fresh fruits and vegetables. Fruits and vegetables provide the essential minerals, vitamins, and fiber required for maintaining health. Fruits and vegetables account for nearly 90% of the total horticulture production in the country. It is the world's second largest producer of fruits and vegetables, after only China and the leader in several horticultural crops, including mango, banana, papaya, cashew nut, areca nut, potato, and okra. According to the National Horticulture Database (Second Advance Estimates) issued by the National Horticulture Board, India produced 102.48 million metric tonnes of fruits and 200.45 million metric tonnes of vegetables for 2020-21. Fruit cultivation covered 9.6 million hectares, while vegetable farming covered 10.86 million hectares.

According to FAO (2021), India is the leading producer of ginger and okra among vegetables, and it ranks second in the production of potatoes, onions, cauliflowers, brinjal, cabbages, and other vegetables. Among fruits, the country produces the most bananas (26.45%), mangoes (including mangosteens and guavas) (43.80%), and papayas (39%). Even though India's share of the global market is only about 1%, its horticultural products are gaining in popularity.

Providing food security to an ever-growing population while ensuring sustainable development is currently one of the most important global challenges. According to the FAO, food production must increase by 70% in order to feed the world's population, which will reach 9 billion by 2050.There is an imperative need for an integrated and innovative approach to the global effort to produce and consume food in a sustainable manner. Huge quantities of food are lost annually on a global scale as a result of post-harvest damages caused by physical factors (mechanical injuries), physiological factors (wilting, shriveling, frost injury), or pathological factors (decay due to fungi and bacteria). In the majority of instances, mechanical damage can increase the likelihood of post-harvest degradation. Post-harvest losses range between 20 and 50 percent in developing nations and 10 to 15 percent in developed nations. Losses in developing nations are unusually high due to a number of factors, such as poor hygiene measures, lack of sufficient storage and marketing facilities, etc. According to estimates, agricultural products worth 580 billion rupees are wasted annually. Even in developed nations, this direction of research and development receives limited attention. In 2010, a study conducted by the University of California revealed that 95% of agricultural research funds were allocated to production enhancement, while only 5% were allocated to the study of post-harvest management and infrastructure. Due to the tremendous post-harvest losses, all efforts to increase crop yields are proving futile. In addition, the starvation problem cannot be resolved solely by advancing agricultural technologies and practices. The urgent need is to discover a solution based on low-cost, locally-available technologies and the best natural and human resources. Even though India is a major producer of horticultural products, most Indians are unable to meet their daily fruit and vegetable demands, resulting in a low HDI-Human Development Index. Due to decay caused by post-harvest pathogens and improper handling, a considerable quantity of fruits and vegetables, approximately 12 million tons of fruits and 21 million tons of vegetables, are lost annually. It is essential to observe that an increase in yield is only meaningful if the produce reaches consumers in good condition and at a reasonable cost. Different employees have periodically proposed a number of strategies for reducing post-harvest losses. Each of these strategies has its own advantages and disadvantages. These strategies are applicable at various phases, including harvesting, processing, sorting, packaging, storage, transport, and marketing, among others. Processing is a vital value-added activity that stabilizes and diversifies the food supply and creates employment and income opportunities. The geographical location of India can be an added advantage for the export of processed fruits and vegetables to stimulate economic growth. Processed foods are more stable, have enhanced nutritional value, and have a longer expiration life. Thus, processing is of great importance in the fruit and vegetable sector, as it not only increases marketing opportunities, but also adds value and reduces post-harvest losses.

Importance of processing industry in India

India is a prosperous agricultural nation, with over fifty percent of its population employed in the agricultural sector. Consequently, India has an excess of raw materials for food processing industries. The total amount of horticultural products produced in India in 2021-2022 was 341.63 million tons. With the assistance of food processing industries, India is able to readily transform raw materials into processed foods. As the agricultural sector is the largest contributor to the Indian economy, all affiliated industries, such as the food processing industry, contribute to job creation and benefit the country's economy. According to Food Processing Ministry of India, the food processing industry contributes 12.8% to India's GDP in 2020-21. The next step in sustainable development is food processing because processed foods are less likely to spoil, thereby reducing food waste. Due to their prolonged shelf life, processed foods can be exported to various countries with relative ease.

Scope of processing industry in India

India has a lot of opportunities for growth in the food processing industry. Because of this, between March 2017 and April 2020, India got about USD 7.54 billion in FDI in the food processing business. India is the number one producer of some of the world's most important fruits and vegetables commodities, such as mango, papaya, banana, milk, ginger, guava, etc. India also makes a lot of sugar, cashew nuts, potatoes, rice, wheat, and other things. Spices are one of the foods that India exports the most. India uses FPIs to make processed spices that can last for more than a year and are easy to move. IBEF (India Brand Equity Foundation) says that in the financial year 2021, India shipped spices worth $4 billion. India makes many different kinds of food throughout the year because it has a wide range of climates and soils. Also, there are a lot of seashores, rivers, and lakes where non-vegetarian food can be found. Setting up FPIs in the rural parts of India gives them access to a lot of resources and cheap labor all in one place. So, from 2014 to 2015, the Indian government gave NABARD a total of Rs. 2,000crore to help the food production industries and food parks. With the help of Food Processing Industries (FPI), Indian farmers will be able to sell their crops straight to the industry, which will help them make more money. With more people, there is a lot more demand for food, which means that more FPIs are needed to meet the demand.

The Indian government is putting in a lot of work to improve FPIs in India, so many purchases in FPIs can be quite profitable. Several factors like availability of cheap labour, Government Subsidy for cold storage and processing units, convenience of roads in case for marketing and transport providing a great opportunity for processing industries in our country. Availability of cans, bottles, and other equipment at cheap rate, there is tremendous scope for export of processed products like Jam, jelly, marmalade, pickles, etc. dehydrated and dried vegetables in addition to domestic demand in India.

Principles of preservation

On the basis of their mode of action, the three main food preservation methods can be categorized as follows:

1) Slowing down or inhibiting chemical deterioration and microbial growth.
2) Directly inactivating bacteria, yeasts, fungi, or enzymes.
3) Preventing recontamination prior to and following processing.

Removal of micro-organisms or their inactivation: This is accomplished by removing air, water (humidity), decreasing or raising the temperature, and increasing the salt, sugar, or acid concentration in foods. To preserve green leafy vegetables, the water must be removed from the leaves so that microorganisms cannot persist. This is achieved by dehydrating the verdant leaves until all moisture has evaporated.

Inactivating enzymes: Enzymes found in foods can be inactivated by altering their environmental conditions, such as temperature and moisture. One method of preserving legumes is to boil them for a few minutes. This technique, also known as blanching, deactivates enzymes and thus aids in food preservation.

Removal of insects, worms and rodents: By storing food in dry, airtight containers, it is possible to prevent insects, worms, and rodents from destroying it.

Constraints for processing industries

Climate: Many farmers in India lose their crops every year because of bad weather. This makes it hard for food processing companies to get the raw materials they need.

Lack of technology: Food processing businesses still need to use more up-to-date technology to make their work more efficient.

Infrastructure: When there isn't enough infrastructure, it's hard to move food from one place to another.

Food Standards: Every country has its own food standards for things that people eat, so the businesses that make food must meet all of these food standards. But India doesn't have enough labs to check that these foods meet the norms.

Commercial processing technology of fruits and vegetables

Canning: This is a process where fruits and vegetables are sealed in airtight containers and heated to a high temperature to kill any bacteria and ensure a long shelf life. Heat is used to remove spoilage organisms in canning, and the food is stored in airtight containers to prevent spoilage organisms from entering. Heat treatment under high pressure in a retort is used to sterilize sealed cans. The temperature required for successful sterilization varies with the pH content of the product and is typically higher than the boiling point of water (100°C or 212°F). It is especially beneficial in developing nations when refrigeration is limited or non-existent. The main concept of canning is to use heat to kill all spoilage organisms inside the sealed container. The containers are sealed while still hot in order to create a vacuum within when they are cooled to room temperature. Canned fruits and vegetables that have been properly processed can be stored at room temperature for years. Some canned fruits and vegetables may have worse quality and nutritious value than fresh and frozen items due to the high heat treatment (Rickman *et al.*, 2007a; 2007b). The inferior product quality compared to aseptic processing could be attributed to the slower heat for a longer period of time. Chemical preservatives like salicylic acid or "canning acids or powders" are unhealthy and should be avoided when canning at home.

Freezing: Freezing is a method of food preservation that involves lowering the temperature to inhibit the development of microorganisms. It is still one of the most effective ways to preserve vital nutrients such as carotenoids in vegetables (Rickman *et al.*, 2007). It is primarily applied to semi-processed fruit products that are intended to be very high quality and expensive (Dauthy, 1995). The invention of the individually quick-freezing (IQF) method has made it possible for preserved foods to be of superior quality. IQF is a technique that prevents the formation of massive ice crystals in vegetable cells. The nutritional value and quality of frozen fruits and vegetables are superior. In the preservation of fruits and vegetables, various freezing techniques are extensively utilized. These methods consist of blast freezing, plate freezing, belt-tunnel freezing, fluidized-bed freezing, cryogenic freezing, and dehydro freezing. Mechanical refrigerators chill products with fluid, air, or surfaces. Commercial freezers use spiral blast freezers. This freezer circulates air at 1.5–6.0 m/sec at −300C to −400C over fruits and vegetables. Cryogenic freezers directly contact products with liquid or solid carbon dioxide or liquid nitrogen (Fellows, 2000). Plate freezers are utilized for prepared rectangular products, while cryogenic freezing is mostly employed for juices. Cantaloupe, figs, fig puree, fruit bar, honey dew, mango, mango puree, melon, pineapple, pineapple puree, leafy greens, mushrooms, okra, peppers, summer squash, sweet potatoes, mushroom and pecan puree, pepper puree are among the numerous types of frozen tropical vegetables and fruits (APEDA, 2007). At 0 degrees Fahrenheit (-18 degrees Celsius), fruits and vegetables can typically maintain their quality for 12 months and 8 to 12 months, respectively. Increasing the storage temperature decreases shelf lives. The storage time is approximately reduced in half for every 18o F (10o C) increase in temperature (Srivastava and Kumar, 2007). Fruits and vegetables freeze below their freezing point. Water activity, enzymatic processes, and microbiological growth decrease. Freezing damages texture of fruits and vegetables. Before freezing, many fruits and vegetables must be blanched to reduce microbial load and inhibit enzymatic activity, including polyphenol oxidase (PPO) browning and lipoxygenase lipid oxidation. Blanching frozen vegetables destroys lipoxygenase, which causes odd aromas, smells, loss of carotene, chlorophylls, and important fatty acids. Blanching targets peroxidase since lipoxygenase is hard to measure. Strawberries, raspberries, spinach, peas, beans, and potatoes are commercially frozen.

Drying: Drying is a process where water is removed from fruits and vegetables to make them shelf-stable. This can be accomplished by means of sun-drying, air-drying, or specialized apparatus such as dehydrators. The purpose of dehydrating is to obtain a solid product with an insufficient amount of water. It is one of the earliest food preservation methods. The main advantage of removing water is the significant reduction in volume and weight, which greatly aids in the storage and transportation of dried materials. Sun drying is the simplest and most straightforward method of dehydration, but the items obtained are of poor quality. For microorganisms and enzymes to activate food deterioration mechanisms, water is required. In this method, the relative humidity is reduced to the point where the microorganisms' activity is inhibited. The majority of microorganisms can thrive at water activities greater than 0.95. When water activity falls below 0.90, bacteria are quiescent. The majority of microorganisms cannot thrive at water activities lower than 0.88. There are numerous advantages to drying. It reduces the weight and volume of foods, makes storage, packaging, and transportation easier, and imparts distinct flavors and aromas. Dried fruit and vegetable products normally contain 1-5% moisture and their shelf life at room temperature may be 1 year or longer. With all these advantages, dehydrating appears to be the least expensive method of food preservation.

Dehydration: Dehydration is the process of removing nearly all of the water from crops under controlled conditions while generating minimal (or ideally no) changes in food qualities (Potter and Hotchkiss, 1995). Dehydration extends the shelf life of fruits and vegetables by reducing water activity, which is the amount of water available for microbial growth and enzyme activity. As a result, removing enough water can limit microbial growth and unwanted enzyme activity. Dehydration involves the removal of water through evaporation, such as sun drying, tunnel drying, and freeze drying. Because of their simplicity and low cost, sun and solar drying are commonly used to dehydrate fruits and vegetables all over the world. However, this approach results in microbial contamination and some quality alterations, such as color and texture. It also necessitates a significant quantity of area as well as a hot, dry climate. Tunnel driers are often used for quickly dehydrating fruits and vegetables. Tunnel driers dry products by stacking trays on trucks that drive constantly through a tunnel. Freeze drying is a technique developed to reduce quality fluctuations. To protect the quality of the product, water in fruits and vegetables is removed by sublimation at low temperature and low pressure; hence, water in the products is transformed into ice first and then directly from the solid into the gaseous state without becoming liquid.

Juice extraction: Fresh fruits and vegetables can be juiced to make a variety of products, including juices, smoothies, and concentrates. It is done by size reduction methods such as crushing, chopping and communiting. Modern methods such as centrifugation, diffusion, pressing and reverse osmosis are used.

Pickling: Pickling is the process of putting food in an acidic solution, usually vinegar, or a salty solution, called brine, to keep it from going bad. In the second case, fermentation makes the acid that does the preserving, which is mostly lactic acid. Pickling is also called chilling, and the food that comes out of it is called a pickle. Most of the time, pickles are made from vegetables, but fish, eggs, and meat can also be pickled. Most of the time, this process is called healing. Most of the vegetables and fruits that are pickled for sale in Western countries are cucumbers, cabbage, and green olives. Other vegetables that can be pickled are capers, garlic, onions, carrots, cauliflower, beans, and so on. There are two kinds of pickled vegetables: those that have been preserved and those that have not. Sauerkraut, which is made from cabbage, and kimchi, which is made from Chinese cabbage or white radish, are both fermented pickled foods that come from lactic acid fermentation. Depending on the acidity of the environment, fermented or pickled vegetables require less heat treatment before being placed in containers. The greatest acidity level obtained is usually around 1.5% lactic acid (and very rarely 2.5%), and it takes 4-6 weeks to achieve. The ideal acidity is 1.0-1.8% with a pH of 4.1 or lower. In the initial phase, the fermentation temperature is 20-25°C and should be reduced to 14-18°C. Fermented cabbage (Sauerkraut) requires maximum acidity level of about 1.5% lactic acid (and very rarely 2.5%); this is obtained in 4-6 weeks. Optimal acidity is 1.0-1.8% and pH value 4.1 or lower. Fermentation temperature is at 20-25°C in the first phase and needs to be lowered then to 14-18°C. During fermentation, the brine from each storage / fermentation silo cell is periodically circulated with a pump in order to make a uniform and stable fermentation process (Dauthy, 1995). The Chinese cabbage, cabbage, tomato, carrot, and spinach prove comparatively more fermentability than other vegetables (okra and gourds) because they have higher fermentable saccharides (Kim *et al.,* 2000).

Irradiation: Food irradiation is a technology that can be used safely to reduce food losses due to deterioration and to prevent contamination that causes illness and mortality. Utilizing electron beams, gamma rays, or x-rays, food irradiation eliminates harmful microorganisms, insects, fungi, and other parasites, and retards spoilage. It does not radioactively alter food. Radiation destroys and renders pathogens incapable of reproduction. Food Irradiation is an umbrella term for multiple processes. The purpose of each process is to eliminate or inhibit the ability of unwanted organisms to reproduce, or to alter the morphology of the product in a manner that extends its shelf life. The optimal dose of ionizing energy (radiation) for each procedure depends on the desired effect. Pasteurization (Pathogen Reduction), Sterilization, and Sanitation are significant processes. This technology is very prevalent in stopping potatoes from sprouting during long-term storage.

Innovative technologies for fruit and vegetable processing

One of the primary responsibilities of the food industry is the production of high-quality fruit/vegetable products with excellent sensory quality, long shelf life, and high nutritional value (Barrett and Lloyd, 2012). To meet the requirements, new processing technologies have been developed. There is potential for introducing new products to the market through innovative techniques, as they have enormous potentials for food processing with the preservation of essential properties, increasing the food's competitiveness to eliminate the use of certain artificial food additives and reducing energy costs. Some innovative technologies are as follows:

Pulsed electric field: Pulsed electric field (PEF) food processing is defined as a procedure in which food is put between two electrodes and exposed to a pulsed high voltage field (20- 40 kV/cm). PEF treatment time is often shorter than one second. This process uses low processing temperature and short residence time allow for highly successful microbial inactivation. Gram-negative bacteria are substantially more effectively destroyed by PEF processing than gram-positive bacteria. Vegetative cells are substantially more vulnerable to this process than spores. The breakdown of cell membrane function and electroporation cause all cell deaths. PEF technology preserves the taste, flavor, and color of meals. This technique is also non-toxic. This procedure, however, has no effect on enzymes or spores. It is also unsuitable for conductive materials and can only treat liquid foods. This procedure consumes a lot of energy and may be hazardous to the environment.

Ultrasonic Processing: The ultrasonic method is one of the more rapidly growing approaches for reducing processing time, improving quality, and ensuring the safety of food items (Knorr *et al.,* 2011). Sonication is a non-thermal technology that uses sound waves with frequencies more than 18 kHz to prepare and preserve food without altering nutritional content (Ravikumar *et al.,* 2017). It can be used for large-scale commercial applications like emulsification, homogenization, extraction, crystallization, dewatering, low-temperature pasteurization, degassing, defoaming, enzyme activation and inactivation, particle size reduction, extrusion, and viscosity modification

(Patist and Darren, 2011). It is utilized to retain both pre- and post-harvest quality features in fresh fruits and vegetables (Gallego-Juárez *et al.*, 2010), and it is also being considered as an alternative for fruit and vegetable cleaning in the food industry (Alexandre *et al.*, 2013). This approach can also detect honey adulteration and perform protein analysis (Awad *et al.*, 2012). Ultrasonication is the most competent method (Abid *et al.*, 2013) for meeting the consumer needs of not only preserving but also enhancing the nutritional value of fruit juices (Bhat *et al.*, 2011a; Bhat *et al.*, 2011b), as this technique retains fresh quality, nutritional value, and microbiological safety in tomato juice (Wu *et al.*, 2008). Orange juice (Valero *et al.*, 2007) and guava juice (Cheng *et al.*, 2007). Ultrasonic cleaners (20-400 kHz) have been used successfully to produce contamination-free fruits and vegetables (Lin and Erel, 1992).

Ozone Processing: Ozone is a highly effective disinfectant since it is the tri-atomic form of oxygen. It is a safe, very effective disinfectant and oxidizer that controls undesired organism development in food processing equipment. It does not act as a systemic toxin on microorganisms, but rather removes them through the oxidation process. A bacterium cannot develop resistance to oxidation since it is 50 times more strong and over 3000 times faster acting than chlorine bleach. Ozone processing in the food sector, either through gaseous treatment or washing with ozonated water, has proven to be effective for fresh fruits and vegetables (Karaca and Velioglu, 2007; Baszak *et al.*, 2019). Spray and fume washing methods can be utilized to minimize microbial counts on the surface of produce (Hampson and Fiori, 1997). Ozone concentrations ranging from 0.005 (control) to 5.0 mol/mol were used to treat tomatoes against *Alternaria alternata* or *Colletotrichum coccode*, with a significant reduction in fungal lesion formation lasting up to 13 days at 13o C (Tzortzakis *et al.*, 2008).

High pressure processing: Fruit and vegetable products that undergo high pressure processing are of superior quality, safer, and have a longer expiration life (Butz and Tauscher, 1998; Gould, 1996; Hayashi, 1992). As covalent bonds are unaffected by pressure, the treatment may be less damaging than thermal processes to low molecular weight food compounds such as flavoring agents, pigments, vitamins, etc. (Tauscher, 1995).

Preparation of different processed products of fruits and vegetables

Preparation of tomato juice: Tomato juice is sterilized in a boiling water bath before being made and stored. Boil the tomato chunks and then strain the pulp through a fine strainer or sieve to remove the seeds and soften the mass. To taste, add a tea spoon of salt or 3-5 g citric acid per litre of juice. The bottling process is the same as for the other juices. Fill the bottles halfway with boiling juice and seal them. Place the bottles and jars in a boiling water bath for 15-20 minutes to heat.

Preparation of fruit syrup: When storage space is limited or vessels are difficult to obtain, fruit syrups can still be made. This technique requires an abundance of sugar. With the majority of fruits, commence with the extracted juice. 2. Bring the juice to a boil and add 1.5 kilograms of sugar per liter of juice. Stir the sugar to dissolve it. If necessary, skim the liquid and then allow it to settle. When using citric acid, dissolve it in hot water before allowing it to settle. Mix the lemon juice or citric acid that has cooled with the syrup, then pour the mixture into the receptacles. Berries, cherries, and plums should be crushed and pressed through a sieve; oranges, grapefruits, etc., should be compressed. While stirring, strain the juice and add lemon juice or citric acid solution to flavor along with 1.5 kilograms of sugar per liter of juice. Remember to agitate the liquid frequently until all of the sugar has dissolved. This may require a day or even more time. Once the sugar has been completely dissolved, pour the syrup into bottles and seal them securely.

Fruit jelly: Jelly is made using fruit juice and sugar. Take out the juice. Apple, grape, red currant, black currant, and elderberry juices are particularly suitable for jelly manufacturing. The following is a general recipe. Boiling reduces the fruit juice to 2/3 of its original volume. Add 3/4 kilogram sugar per litre of reduced juice while stirring. If desired, add lemon juice or citric acid. Boil the jelly mass until a few drops have the thickness of jelly when dropped onto a plate and cooled. Remove any foam. Fill clean jars halfway with jelly and shut immediately with cellophane, a metal, glass, or plastic lid, or greaseproof paper. The jelly can alternatively be covered with hot paraffin wax, which must be followed by a second layer to completely seal all sides. Another jelly recipe that uses less fuel but more sugar is as follows: bring one litre of juice to a boil and add 1.5 kg of sugar. 5 minutes on the stove. Fill the jars and close them as directed above. Jelly can also be created with pectin (follow the directions on the pectin container) or albedo (orange peel white).

Candied fruit: Candying involves gently impregnating the fruit with sugar until the sugar concentration is quite high, approximately 65-70%. Peel and cut the fruit into 1-2 cm thick slices. Boil these pieces in water until a fork can easily pierce them. Soak them in a 30% sugar solution overnight. Following that, the sugar solution is increased by 10%, and the mass is briefly brought back to a boil before being allowed to stand overnight. This procedure is repeated until the sugar solution comprises 72% sugar or less. A sugar refractometer, a tiny, portable, and inexpensive tool, can be used to determine the sugar concentration. Keep the fruits in this +/- 72% saturated sugar solution for many weeks before drying. To avoid crystallization, the sugar solution must contain both glucose and beet or cane sugar. If this is unavailable, inverted sugar can be substituted. Boil a concentrated solution of beet or cane sugar for 20 minutes with a generous splash of acid (vinegar, lemon juice, citric acid, hydrochloric acid, etc.) to make this.

Jam Preparation: There are two methods for making jam.

1) **Volume reduction method:** The fruit is peeled and cut into large segments. In a lidded pan, cook the fruit with a small amount of water until soft. Puree the produce. Reduce the fruit's volume by two-thirds by boiling it in an uncovered pan. Gradually incorporate 3/4 kilograms of sugar per kilogram of fresh fruit into the fruit mélange and continue to boil for several minutes. One can add lemon juice or citric acid to increase acidity. Boil the jam until a few droplets, when sprinkled on a plate and allowed to cool, have the consistency of jam. If required, skim the substance. Fill the vessels according to the instructions for fruit jelly. If you use a lid that can withstand high temperatures, invert the jars so the hot jam will destroy any microorganisms on the lid.

2) **Pectin method:** Pectin is a gelatinizing agent that is used to set marmalade. Follow the instructions included with the product. Pectin can be substituted with apple pulp (apple sauce) or ground al-bedo (the white of orange rind).

Preparation of chutney and marmalade: Use 1 dl of vinegar (5%), 125 g of brown sugar, shallots, Spanish peppers, ginger powder, and mustard powder to flavor 1 kilogram of fruit (tomatoes, rhubarb, etc.). Combine all ingredients and simmer until thickened. Complete the preparations using the jam-making recipe. Marmalade is prepared from citrus fruit. Pectin is not required if the epidermis is utilized, which is also possible. As a preservative, sodium benzoate can be added to preserve marmalade for a long time. Use no more than 250 mg per kilogram of marmalade.

Chutney, Sauces and Ketchups

Chutney, sauces, and ketchups are key export products with considerable customer demand. Fruit chutney is a product manufactured in the same way as jam, except that spices, salt, vinegar, and other ingredients are added to improve taste, palatability, and enticing character. It is also a preservative. The final product must contain at least 40% fruit/vegetable juice and 50% total solids. Cooking time and temperature have a great impact on chutney quality, and the best chutney is obtained by cooking for a long time at a temperature below boiling point. To achieve appropriate thickening, cooking is done without a lid, despite the loss of volatile oils from the spices. Chutneys made from tamarind, mango, papaya, tomato, aonla, and other fruits are popular among Indian customers. Sauces (TSS minimum 28%) have a thinner consistency and contain more total solids (minimum 30%) than ketchups. Sauces are effectively made from tomatoes, plums, soybeans, apples, papayas, and mushrooms, among other fruits and vegetables. Sauces of high quality are made by macerating spices, herbs, fruits, and vegetables in cold vinegar or by boiling. Thickening agents can also be added to sauces to prevent solid particles from settling. Sauces are made in much the same way as chutney, except that the fruit/vegetable pulp or juice is sieved after cooking to remove the skin, seeds, stalk of fruits/vegetables, and spices, resulting in a fine/smooth end product.

Types of fruit beverages: Unfermented and fermented beverages

A. Unfermented beverages

Unfermented beverages are fruit juices that do not undergo alcoholic fermentation. Natural fruit juices, sweetened ready-to-drink drinks, nectar, cordial, squash, crush, syrup, fruit juice concentrate, and fruit juice powder are among them. Tropical and subtropical fruits and vegetables have the potential to be used in the development of value-added beverages loaded with phyto-chemicals that can quench people's thirst in hot and humid weather (Chakraborty *et al.,* 2010; 2011). For preparation, the minimum standards established by the Food Safety and Standards Regulation, 2011 and the Fruit Products Order are observed.

Fruit juices are generally consumed in either fresh or processed form. Certain vegetable juices, such as bottle gourd (Majumder *et al.,* 2011) and bitter gourd (Satkar *et al.,* 2013), are also used in fresh or processed form for medicinal purposes. Tomato juice is commonly consumed in canned or bottled form. Tomato is often used in soup and as a warm beverage before meals.

B. Fermented beverages

These are fruit juices that have been alcoholic fermented by yeasts such as *Saccharomyces cerevisiae*. The amount of ethyl alcohol produced by the product varies. Grape, custard apple, jamun, muskmelon, coconut toddy, pomegranate, banana, guava, ber, strawberry, peach, kiwi fruit, raspberry, cherry, pineapple, date, apricot, litchi and mixed fruits, mango, jack fruit, beet root, watermelon, and others are suitable for winemaking (Soibam *et al.,* 2014; Baidya *et al.,* 2014). Fermentation is a potential process for creating new food items with altered physicochemical and sensory properties, including flavor and nutritional components. Alcohol, acetic acid, and lactic acid are vital in fermented food and beverage production for imparting quality (Ray and Joshi, 2014). Different types of table wines, such as dry, sweet, and sparkling wines with an alcohol content of 8 to 15% by volume, are popular around the world. Wine has significant health advantages in the prevention and treatment of cardiovascular disease (Saranraj and Ray, 2019).

Preparation of RTS

Ready-to-serve (RTS) fruit drinks contain at least 10% fruit juice and 10 degree Brix total soluble solids (TSS) along with 0.3% acidity. It is a ready-to-serve beverage because it is not diluted before dispensing. Mango drink, guava drink, and pineapple drink are commercially available products.

Sweetened fruit juice is classified as a ready-to-drink beverage with a minimum juice content of 85% and a minimum TSS of 10%, whereas unsweetened juice is a natural juice consisting of 100 percent fruit juice with natural total soluble solids. These are preserved through thermal processing. Apple and pear liquids with added sugar are commercially produced. Additionally, RTS beverages can be preserved by adding no more than 70 ppm SO_2 or 120 ppm benzoic acid.

Processing

Mix fruit pulp or liquid with a syrup made by combining sugar, water, and citric acid. Homogenize the mixture to ensure adequate blending. To pasteurize the beverage, put it to a boil. Fill the 200ml glass vessels while they are still hot. Crown the bottles' corks and heat them for 20 to 25 minutes. After the bottles have cooled, label them and store them in a cool, dry area. As a preservative, potassium meta-bisulphite (KMS) or benzoic acid can be added to the RTS drink. When declared on the label, synthetic and authorized artificial color and flavor can be added.

Preparation of Squash

Squash is a fruit-flavored, non-alcoholic concentrated syrup produced from fruit juice, water, sugar, or sugar substitute. Elderflower and ginger extracts are common in traditional squashes. Squash must be diluted before consuming. Squash usually starts with citrus fruits like orange, lime, and lemon or a mix of fruits and berries.

As per FSSAI specifications squash to have 25% fruit content and 40° Brix total soluble solids. Squash should not exceed 3.5 percent anhydrous citric acid. 350ppm of sulphur dioxide or 600ppm of benzoic acid is the squash preservative limit. Potassium metabisulphite bleaches anthocyanin colors in dark fruits. Sodium benzoate is used.

Processing

Selecting ripe fruit. Unripe and overripe fruits degrade juice.

Washing: Dirt and spray residues of arsenic, lead, etc. are removed by washing with water or dilute hydrochloric acid (0.5%) followed by water.

Sorting and grading: Use a table or machine to hand inspect ripening fruits.

Peeling/deseeding/destoning: The selected fruits undergo peeling, de-seeding, and de-stoning.

Cutting/slicing/crushing: Use a slicer to slice and smash the peeled fruit. The juice extraction mechanism depends on the fruit. Citrus fruits, which are inherently juicy, are pressed mechanically. Mango, guava, pineapple, and strawberry juice is extracted by pressing pulp through a perforated metal plate.

Pasteurization: After extraction, juice is pasteurized at 80-95°C for 10-12 minutes. Plate pasteurizers are the most common juice pasteurizers.

Clarification: Fast procedures like centrifugation (6000–6500 RPM). Filtration makes juice transparent. For clear juice, a continuous or decanting centrifuge with automatic desludging works well. Utilize ultra-clarifier.

Sugar, water, and acid are cooked together to make syrup.

Tank mixing: Syrup and clarified juice.

Preservative: Usually sodium benzoate or KMS. Sodium benzoate uses 1.0gm/lit.

Products are cooled to room temperature.

Deaeration: Oxygen in freshly extracted juice can degrade its quality if not eliminated before packing.

Bottle and cap. Aseptic processing and packaging involves packing a commercially sterile product into a pre-sterilized container in a sterile environment.

Conclusion

Fruits and vegetables are consumed primarily for their nutritional value. Despite massive production of fruits and vegetables, a large portion of the population is unable to obtain a sufficient amount of their balanced diet due to a variety of problems. A large population size, poverty, ignorance, and other factors may be at fault. However, one of the main reasons is post-harvest deterioration, which results in a loss of both quality and quantity. A range of initiatives are made to reduce these losses through the use of appropriate post-harvest management measures. Cleaning, washing, disinfecting, drying, packaging, and storing are all examples. Produce is often prepared using various procedures to increase shelf life and add value. Fruits and vegetables have been processed since the beginning of time. It extends the availability of many seasonal fruits and vegetables throughout the year. It also aids in the reduction of post-harvest losses and the addition of value. However, consumer worry about the safety of the food, combined with its higher quality, necessitates the development of better solutions to address the issue. There is a need to develop more efficient processing processes to increase crop value while minimizing post-harvest losses. This would also assist underdeveloped countries maintain economic stability.

Countries such as India may also benefit from their connectedness to the majority of the world for export or import purposes. However, technology advancement, quality management, and strict adherence to export obligations are required to maintain global standards. The processing of fruits and vegetables sector in India has a large scope with promising future prospects, but it also faces a number of problems in competing globally.

References

1. A Montano A, Sanchez A.H, Beato V.M, Lopez A, De Castro A. 2016. Pickling. *Elsevier,* 369-374.

2. Amit S.K, Uddin M.M, Rahman R, Islam S.M.R, Khan M.S. 2017. A review on mechanisms and commercial aspects of food preservation and processing. *Agric. & Food Secur.* 6-51.

3. Carl L.L. (1940). Canning Fruits and Vegetables. 421, 1-9.

4. Chakraborty I, Lalramhlimi B, Saurabh V. Processing and value addition of fruits and vegetables. Postharvest Management of Horticultural Crops. Jaya Publishing House, New Delhi. 2021; 1, 608-676.

5. Chauhan J, Goswami T.K, Mangaraj S, Nath A. Post-harvest management and production of important horticultural crops. Horticultural crops processing. *Scientific Publishers*, New Delhi, 2016; 1, 210-275.

6. Chemat F, Rombaut N, Meullemiestre A, Turk M, Perino S, Tixier A.S. F, Vian M.A. Review of Green Food Processing techniques. Preservation, transformation, and extraction. Innovative Food Science and Emerging Technologies, 2017; 41, 357-377.

7. De-Corcuera J.I.R, Schneider R.M.G, Barringer S.A, Urbina M.A.L. Fruits and Vegetables - Processing Technologies and Applications Food

Processing: Principles and Applications. John Wiley & Sons, Ltd. John Wiley & Sons, Ltd. 2014; 2, 339-362.

8. Geetanjli Scope and challenges in fruit and vegetable processing for reduction of post-harvest losses and value addition. *International journal of advance research in science and engineering*, 2017; 6(3), 346-353.

9. Holdsworth S.D. The Preservation of Fruit and Vegetable Food Products. 1983; 2, 9-11.

10. https://apeda.gov.in/apeda

11. https://www.energyreview.in/post/a-sustainable-solution-for-post-harvest-management-in-india

12. Jadhav H.B, Annapure U.S, Deshmukh R.R. Non-thermal Technologies for Food Processing. *Front. Nutr.* 2021. 8, 657090. DOI: 10.3389/fnut.2021.657090.

13. James I.F, Kuipers B. Preservation of fruit and vegetables. Agromisa Foundation, Wageningen, 2003; 4, 6-68.

14. Jana B.L. (2012). Postharvest technology of agri-horticultural crops. Processing and preservation of fruits and vegetables. *Pointer publishers*, Jaipur, rajsthan, 1, 205-222.

15. Jongen, W. (2002). Fruit and vegetable processing Improving quality. Minimal processing of fresh fruits and vegetables. Wood head Publishing Ltd the authors have asserted their moral. 1, 288-306.

16. Kalyani B, Manjula K. 2014. Food Irradiation- Technology and Application. *Int. J. Curr. Microbiol. App. Sci*, 3(4), 549-555.

17. Montano A, Sanchez A.H, Beato V.M, Lopez L.A, Castro A. 2016. Pickling. *Encyclopedia of Food and Health*, 10.1016/B978-0-12-384947-2.00545-6.

18. Peter K.V. (2007). Post-harvest technology of horticultural crops. *New India Publishing Agency*, Pusa New Delhi, 1, 1-119.

19. Sumonsiri N, Barringer S.A. 2014. Fruits and Vegetables-Processing Technologies and Applications Food Processing: Principles and Applications. John Wiley & Sons, Ltd. John Wiley & Sons, Ltd. 2, 363-381.

20. Teixeira A.A. Thermal Food Preservation Techniques (Pasteurization, Sterilization, Canning and Blanching). *John Wiley & Sons*, Ltd., 2015; 1, 115-128.

21. Tylewicz U, Tappi S, Nowacka M, Wiktor A. Safety, quality, and processing of fruits and vegetables. *Foods*, 2019; 8, 569.

Chapter - 2

Preparation of Jam/Jelly/Preserve/Candy, Preparation of Toffee/Bar/Cheese

Authors

Vaishnavi

Ph.D. Scholar, Department of Food Science, Nutrition and Technology, CSK Himachal Pradesh Agriculture University, Palampur, Himachal Pradesh, India

Mamta

Ph.D. Scholar, Department of Food Science, Nutrition and Technology, CSK Himachal Pradesh Agriculture University, Palampur, Himachal Pradesh, India

Nivedhika Rai Vaid

Assistant Professor, Chandigarh College of Hospitality, CGC, Landran, Punjab, India

Pooja Bhatt

Assistant Professor, Chandigarh College of Hospitality, CGC, Landran, Punjab, India

Chapter - 2

Preparation of Jam/Jelly/Preserve/Candy, Preparation of Toffee/Bar/Cheese

Vaishnavi, Mamta, Nivedhika Rai Vaid and Pooja Bhatt

Abstract

This chapter shed light on the preparation of jam, jellies, cheese, bar and other preserves. Production of jam, jellies and other preserves is an ancient and widely practiced technique for preserving fruits and vegetables for much longer than their typical storage life and making use of blemished or off-grade fruits that may not be ideal for fresh consumption. People can still enjoy their favorite fruits even in rough times or when fresh food is expensive. To use processing and preservation methods at both the farmer and industrial scales to avoid losing the seasonal surplus of fruit should be warranted. Such efforts will help the development of processing industries in the growing areas of the countries.

Keywords: Fruits, jam, jelly, preserve, bar, cook

Introduction

A diversity in agro-climatic conditions across India allow cultivation of numerous horticulture crops. India is second-largest producer of fruits and vegetables. Additionally, nearly 30% of the world's fruit production comes from India. The country has diverse topography and climate, which ensures availability of fruits and vegetables in every season. Fruits contain a lot of sugar and are a good source of vitamins A and C. Fruits and vegetables plays a critical role in human diet and nutrition. About 16 to 72 percent of fruits are wasted due to due to the perishable nature and season availability, inadequate storage conditions and transportation conditions. Fruit preservation is therefore necessary to decrease fruit waste (Tiwari *et al.*, 2021). Jam, jellies, preserves, and other similar goods are usually made from fruits and vegetables which not only reduce fruits wastage but also allow people to enjoy their favorite fruits even during hard times or when fresh food is costly. Additionally, these goods enable fruit usage diversification, reducing post-harvest waste and avoiding distress sales during the crop season. The making of jellies and other preserves is an old and popular process, providing a means of keeping fruits far beyond their normal storage life and sometimes making use of blemished or off-grade fruits that may not be ideal for fresh consumption. If sufficient food safety-related precautions are taken, these techniques for producing the processed products could be tried at the household level even in remote areas.

JAM

The fruit jam has a smooth consistency, which is produced by crushing fruit and disintegrating the solid chunky leftovers. For the preparing fruit jam the fruit juice is mixed with pectin and heated to form the gelatinous spread containing 65% sugar (Biswas, 2021). Jam is produced by cooking fruit pulp with enough sugar until it is effectively thick and hard for keeping the fruit tissues in place. Jams are made with apples, pears, apricots, loquats, peaches, papayas, carrots, plums, strawberries, raspberries, mangoes, tomatoes, grapes, and muskmelon (Srivastava & Kumar, 2006).

It can be produced using a single type of fruit or from many different kinds. Commercial jams like tutti-frutti can be made from fruit scrapings, fruit fragments, and pulp stuck to fruit cores, all of which are readily available in canning factories. For every 45% fruit pulp, 55% of sugar should be utilized. Jam must have 68.5% TSS, and 0.5-0.6% acid (citric acid) per 100 g of the processed product to meet FPO specifications (Kumar *et al.*, 2017).

Selection of fruit and preparation of jam

a) Pick ripen fruits of good quality. In cold water, wash the fruits. Fruits should be peeled, and any corers and stones should be removed. Using a stainless steel knife, chop up the peeled fruit into little pieces. Hard fruits should be chopped into extremely tiny pieces. Utilizing a pulper, pulp the fruits.

b) Add sugar and acid.

c) Cook mixture by stirring occasionally and with the help of ladle crush food pulp. The bulk should be cooked until it reaches 105.5° C.

d) Flake Test or Sheet: When the jam is boiling is scooped out with a spoon and cooled. Then it is allowed to fall. The end point of finished product is, if it falls off in form of a sheet or flakes rather than flowing continuously like syrup. Boiling should be done if the sheet test comes negative.

e) Packaging: Store jam in a sterilized container with tight lid in a cool temperature (Srivastava & Kumar, 2006).

Process: Select ripe firm fruits → Washing of fruits → Peeling of fruits →Pulping (Remove seed and core) → Addition of sugar and acid → Boiling (with continuous stirring) → Judging of end point by further cooking up to 105° C (or) 68% TSS (or) by sheet test → Transfer in hot sterilized bottles → Capping → Cooling (at room temperature) → Stacking →Labeling → Dispatching (Srivastava & Kumar, 2006).

In home jam can be prepared by using the recipes as given in the table

Fruit/ Vegetable	Ingredient for one kg pulp		
	Sugar (kg)	Citric acid (g)	Water (ml)
Aonla	0.75	-	150
Apple	0.75	2.0	100
Apricot	0.60	1.0	100
Carrot	0.75	2.5	200
Grapes	0.70	1.0	50
Guava	0.75	2.5	150
Karonda	0.80	-	100
Loquat	0.75	1.0	100
Mango	0.75	1.5	50
Musk melon	0.75	2.5	50
Plum	0.80	-	150
Peach	0.80	3.0	100
Pear	0.75	1.5	100

Problem in jam production

1) **Crystallization:** Invert sugar should constitute 30-50% of the finished product. If the proportion is less than 30, can sugar may crystallize during storage, and if it is greater than 50, the jam may form into a honey-like mass as a result of the formation of tiny glucose crystals. Corn syrup or glucose maybe added along with cane sugar to prevent crystallization.

2) **Sticky or gummy jam:** Jams frequently turn gummy or sticky due to the high percentage of total soluble solids. By adding pectin, citric acid, or both, this issue can be resolved.

3) **Premature setting:** This occurs due to jam's low total soluble solids and high pectin concentration, and it can be avoided by increasing the sugar content. If this is not applicable, little sodium bicarbonate is added to lower the acidity and avoid pre-coagulation.

JELLY

When making jelly, a fruit is crushed, and the solid, chunky remains are discarded. Jelly has the smoothest consistency. The only thing left is fruit juice, which is mixed with pectin and cooked to create the 65% sugar gelatious spread. The process of making jelly involves boiling a clear, filtered solution of fruit extracts with pectin that is free of pulp, with the addition of sugar and acid. A jelly should have a translucent consistency, be well-set but not too hard, and maintain the flavour of the fruit from which it was produced. When taken from the moulds, it should maintain its shape and have an appealing colour. It shouldn't have crystallized sugar, sticky, or syrupy. The item must not be tough or rubbery, and it shouldn't have any dullness or syneresis (weeping) (Mondhe *et al.,* 2018). According to FPO specifications the final jelly product must have 65% solids, 45% fruit extract, and 0.5–0.75% acid. Pectin-rich fruits and vegetables such guavas, sour apples, plums, wood apples, papayas, and jackfruits are frequently used for preparing jellies. You can add pectin powder to fruits with low pectin levels, such as to pineapple, strawberries, grapes, etc. The process for making jelly is the same as for making jam.

Process: Selection of firm fruit → Washing and Cleaning → Cutting fruits into thin slices → Boiling with water (1 ½ times the weight of fruits for about 20-30 min) →During boiling addition of citric acid (2 g per kg of fruit) → Continues straining of extract → Pectin test (for addition of sugar) → Addition of sugar → Boiling → Evaluation of end point (sheet / drop / temp test) → Discarding of scum (or) foam (one teaspoonful of edible oil added for 45 kg sugar) → Addition of colour and remaining citric acid → Transfer into sterilized container → Waxing (paraffin wax) → Capping → Storage at ambient temperature

In home jelly can be prepared by using the recipes as given in the table:

Fruit	**Ingredient for one kg pulp**	
	Sugar (kg)	**Citric acid (g)**
Guava	0.75	3.0
Sour apple	0.75-1.00	2.0
Gooseberry	0.80	-
Karonda	0.75	-
Jamun	0.75	1.0
Wood apple	1.00	-
Plum	0.75	2.5
Loquat	0.80	2.0
Papaya	0.75	3.0

Problems in jelly making

1. **Failure to set:** It could be because of

   i) **Addition of excessive sugar:** It results in a jelly that is syrupy or very soft, which can be fixed by adding a suitable amount of fresh, filtered extract that is high in pectin.

   ii) **Lack of acid or pectin:** Lack of acid or pectin, or both in fruit used or insufficient heating of the fruit slices, which limits appropriate extraction of pectin and acid.

   iii) **Heating beyond the end- point:** The jelly will stay syrupy and very spongy if heating is stopped before the total soluble solids percentage reaches 65.

   iv) **Long-term heating:** Prolonged heating should be avoided since, in the presence of acid, pectin loses its ability to coagulate if it is heated for a long period.

2. **Cloudy or foggy jellies:** It could be because of

   i) Use of juice or extract that hasn't been clarified.

   ii) **Use of immature fruits:** Utilizing green immature fruits gives juice a cloudy appearance because they contain starch that is insoluble in juice.

   iii) **Over-cooking:** Due to their high viscosity, such jellies are gummy or sticky and do not clear up after being poured into a container.

   iv) **Scum not being removed:** The jelly will fog up if the scum is not removed before pouring.

   v) **Improper pouring:** When jelly is poured into containers from a considerable height, some air becomes trapped in the form of bubbles and turns the jelly opaque. As a result, the pouring vessel shouldn't be held more than 2.5 cm away from the container's top.

3. **Formation of crystals:** It is caused by the addition of too much sugar and by the jelly being concentrated too much.

4. **Jelly weeping or syneresis:** Weeping, also known as syneresis, is the spontaneous exudation of fluid from a gel and is caused by several factors:

   i) **Too much acid:** This damages the structure of the jelly by causing pectin to hydrolyze or decompose. It is more common in tender jelly and can be preserved through the addition of a high amount of sugar with either more pectin or low-acid juice to increase the volume of the jelly and lessen its acidity.

ii) **Insufficient pectin:** Results in the formation of a pectin network that is not dense and rigid enough to support the sugar syrup.

iii) **Immature gelation:** This results in the jelly being weak and remaining broken as the pectin network breaks during the jelly's pouring into the containers.

iv) **Fermentation:** Although jelly has a high sugar content (65%), normal fermentation can still occur if synersis takes place. Even when sealed with a paraffin wax seal, jelly stored in a moist environment promotes the growth of mould. The growth could be caused by a number of things, including improper jelly covering, insufficient application of hot paraffin wax to kill moulds and microbes, and improper baking of the paraffin wax seal. To overcome this problem, hermetically sealable glass jars and cans are adopted.

Preserve

Fruit preserves is prepared by boiling whole mature fruit or in form of big pieces in a lot of sugar syrup until it is soft and translucent. You can make preserves using different kind of fruits E.g. aonla, mango, apple, strawberry, pear, cherry, karonda, pineapple, papaya. When making preserves, at least 45 kg of fruit must be used for every 55 kg of sugar, and cooking must continue until at least 68 percent of the soluble solid is achieved. FPO specifications for preserves are TSS 68° Bx and 55% fruit pulp.

General considerations: Fruit that has been cooked in syrup directly shrinks and absorbs less sugar. Therefore, before steeping in syrup, in order to make the fruit soft enough to absorb water, it should be blanched. Fruits with a lot of juice, however, can be cooked right away. There are three ways to cook fruit in syrup, which are discussed below:

i) **Rapid process:** Fruits are boiled in syrup using little sugar. The syrup is allowed to boil with low temperatures until it thickens enough. Soft fruits like strawberries and grapes only need to be slightly boiled to soften. Unlike hard fruits that require extensive cooking, including peaches, pear and apple. But avoid quick boiling because it makes the fruit hard especially when cooking in a huge, shallow pan with little amount of syrup. A final sugar concentration should not be below 68%, or a boiling point of 106° must be achieved. Despite being a fast and affordable process, it has drawback that boiling significantly reduces the product's flavour and colour.

ii) **Slow process:** Fruit is softened via. Being blanched. After that, alternate layers of sugar equal to the weight of the fruit is layered in the blend and it is left to stand for 24 hours. During this period, the fruit releases water and the sugar dissolves, resulting in the production of syrup with a TSS level between 37 and 38%. The produced syrup is heated on the next day after the fruits are removed to increase its strength to about 60% TSS. To avoid crystallization, citric acid (1 to 1.5 g/kg sugar) is also added. The entire mass is kept 24 after boiling for 4-5 minutes. By boiling the syrup on the third day, the strength rises to about 65% TSS. Finally, the syrup's strength is increased to 70% TSS and the fruits are allowed to soak in it for a period of one week. After that preserve is ready and can be is stored in sterilized containers.

iii) **Vacuum method:** The fruit is first boiled to soften it, and then it is placed in a syrup that should contain 30- 35% TSS. After that, the fruit-syrup mixture is moved to a vacuum pan and concentrated at a lower pressure to 70% TSS. Compared to the other two processes, this method maintains the fruit's flavour and colour. In each of these procedures, the fruit must be kept submerged in syrup both during and after cooking to prevent drying out and lowering the quality of the final product. After the last boil, the product needs to be quickly chilled to avoid discoloration during storage. The sterilized jar is then filled with freshly made boiling syrup that contains 68% TSS and sealed tightly.

Process: Mature fruits → Washing → Preparation of fruit for sugar treatment → Keeping fruit and sugar in alternate layers (1.0 kg Fruit: 1 kg Sugar) (or) steeping fruit in syrup of 40% TSS for a day → Removal of fruit → Increasing consistency of syrup to 60% TSS by boiling → Steeping of fruit for a day →Repeating the process and raising strength of syrup by 5% TSS to 70% on alternate days – Steeping in 70% TSS for a week → Preserve – Draining – Filling in jar → Covering fruit with freshly prepared sugar syrup of 68% TSS →Sealing (airtight) →Storage

Candies

Candy is a general term for a variety of confections including hard candy, soft candy, taffy, marshmallows and caramels. Candy's key characteristic attribute is the significant quantity of sugar or sugar substitutes that are used in its manufacturing. Candy provides food energy, but often it doesn't provide any additional nutritional value. It is considered as a source of empty calories. Candy has remained popular among people of all ages and is particularly liked by children. Value-added candy with beneficial properties is in high demand on the worldwide marketplace. It has a bright future in the fast food sector because of its "portability and long shelf life without refrigeration". Fruit and vegetable-based candies are becoming more and more common because of their wide acceptance, low volume, increased nutritional content and longer shelf life. The process of producing candied fruit is quite similar to that of making preserves. The fruit is just impregnated with syrup that has a higher concentration of sugar or glucose, which is the only difference. Sugar, acids, and fruit or vegetable pulp are the key elements of candies. Candy can be produced from a variety of fruit and vegetable by-products. Suitable fruit and vegetable used to prepare candies and toffees are Pineapple, Banana, Mango, Papaya, Apple, Guava, Peach, fig, citrus, kinnow, kagzi lime, Tamarind, strawberry, coconut, passion fruit, annona, bael, aonla, Carrot, ginger, ash gourd, watermelon, muskmelon, pumpkin apples, ginger, mangoes, guava, carrots and citrus peels. The method of making candies and toffees involves osmotic dehydration, in which sugar goes into the tissues and allows water to be released into the syrup (Kumar *et al.,* 2017).

Types of candies and toffees

1) **Hard candy (Non-crystalline):** These are a sub cooled, supersaturated sugar solution. They are often produced from sucrose, glucose syrup (or corn syrup), and/or other sugars and sweeteners with a small quantity of flavouring, colouring, and organic acids. Commercial hard candy and toffee preparations have a maximum moisture content of 5% wt/wt. It is uniform and might be chewy or hard.
2) **Soft candy (Crystalline):** They are easily chewable or melt in the mouth due to their creamy, small-crystal structure (Durrani *et al.,* 2011).

Preparation of different candies

Ingredients: Peeled slices-500g, Sugar-750g, Water-625ml, Citric acid-1.5g, Lime-20g

Method

1) Pick fruits that are fully ripe and offer them a thorough wash under running water.

2) Peel off the fruit skin and cut it into the desired shape and size of slices.
3) Soak the slices in the lime water for a full hour.
4) Drain the lime water from the fruits and afterwards give them a good wash to remove all traces of lime.
5) Blanch the pieces for three minutes followed by draining off the water.
6) After three minutes of blanching, drain off the water.
7) To remove extra moisture from the fruit, spread the pieces out on a clean tray.
8) To make sugar syrup, mix 400 grams of sugar with 400 milliliters of water. Place the pieces in the deep container, add the syrup, and keep it for 24 hours.
9) Next day, boil the fruit pieces in the syrup and increase the concentration by 10° B.
10) Repeat this process until the syrup strength reaches 60° B. Add citric acid at this point.
11) Increase the strength of the syrup to 68° B and soak the fruit pieces for a week.
12) Drain the sugar syrup on the eighth day, and then dry the guava fruit pieces at 55 to 60° C for 5 to 6 hours.
13) Place the dried fruit candy in glass jars after coating it with grinded sugar.

Toffee

Butter and sugar or molasses are caramelized to produce toffee. This mixture may contain additional ingredients including nuts, raisins, and flour. The hard crack stage in the production of confections is attained when the above-mentioned combo is cooked until it reaches temperatures of 300 to 310° F. The toffee mixture will have a glossy surface at the hard crack stage, but it will also be rigid enough to be shaped into various different shapes. After that, the mixture is placed into a tray and allowed for cooling and hardening. This slab can be divided into substantial pieces and served as sweets. To create varied flavors, it can also be combined with various components like chocolate, vanilla, honeycomb, raspberry, raisins, etc. Toffee's colour is brown as a result of caramelization.

Structure of toffee

Toffee is an emulsion of fat in complex aqueous system. Water, proteins, and carbohydrates work together to prevent crystallization. Toffees have the following textural characteristics:

- Moisture content has an impact on hardness.
- Carbohydrate molecular weight has an impact on chew ability and toughness when consumed.
- The quantity, level of emulsification, and hardness of the fat, as well as the quality and quantity of milk protein, all affect the body.

By taking into account all these aspects of all the different components that define the toffee's eating capabilities; it is possible to obtain the desired toffee texture through modifying the quantities of the ingredients and using alternative ingredients.

Formulation

Three parts sugar, five parts glucose syrup, three parts sweetened condensed milk, and one-part fat tend to be found in a balanced recipe. For wrapped toffees, the ratio of sugar to non-crystallizing sugars should be 1.1:1.

Processing

Toffees can be produced using equipment ranging from a little saucepan to a large continuous plant. Practically similar processes are used.

a) **Dissolving:** First, water or a solution of water and glucose syrup is used to dissolve solid sugars.

b) **Emulsifying:** For the formation of the emulsion, the fat and skim milk solid are mixed with the dissolved sugar or glucose syrup.

c) **Cooking:** The final water content of the emulsion is then obtained by cooking it. The boiling points of normal toffee and hard toffee, respectively, are 118-132°C and 146-154°C.

d) **Shaping the Toffee:** The slab process, cut and wrap processing, and depositing are the three methods used to shape toffees for individual twist wrapping. Typically, toffee is poured into trays, cut into slabs, or used as an ingredient for other candies.

e) **Slab Process:** This is an actually antique way to shape toffee. Over slabs that have been chilled by water, toffee is gradually poured. The toffee slabs are coated with a releasing agent on them. This used to be mineral oil, but the use of long-lasting vegetable fats has made it illegal. Toffee is typically carefully poured over slabs that have been chilled by water. Because toffee is a poor heat conductor, it must be turned over on the slab to transfer the heat from the warmed center to the cooler outer section. If the toffee was simply left to cool on the slab without turning, the outside would harden before the inside had cooled. After the toffee has sufficiently cooled, it is cut into sheets rather than single pieces.

f) **Cut and Wrap Process:** In this stage, the toffee is cooled, usually by pouring it over a chilled metal band or drum, but if necessary, cooling could also be done over a slab. The cooling temperature must not drop to the level where condensation takes place. Following cooling, the product is fed into batch rollers, which form the mass initially into a cone and then a rope. Later, the rope moves through more rollers to decrease its thickness before passing through forming wheels that produce the desired cross section. A revolving knife is used to cut the product into pieces, which are then fed onto the wrapping paper and twist wrapped. Modern machinery operates at fast speeds. Although it is very quick, this type of system has the drawback that the toffee pieces get distorted during wrapping.

g) **Depositing:** This method is most recent. Soft caramels can be placed in chocolate shells or starch molds, and the chocolate shells or molds can then be sealed by adding chocolate back. The high-tech approach involves depositing the toffee into moulds made of rubber. Making such a deposit is a difficulty since the depositing machinery has to be kept hots enough for the toffee to flow without losing any more water. If kept at a high temperature for a long time, toffee can thicken. The rate at which heat can be removed from the finished toffee is controlled by the thermal conductivity of the toffee, which is low. When the toffees have cooled, they are fed into the packaging machine.

h) **Packaging and storage**: The majority of toffees are packaged in individual wraps composed of waxed paper, aluminium foil, cellulose film, or a combo of these types of materials. Toffees are typically wrapped by hand, but semi-automatic wrapping instruments are available for more production. The individually wrapped candies may be placed in a heat-sealed polythene bag for extra safety. Toffees are mostly stored under the ambient conditions of temperature and humidity.

Bar

For several years now, due to changes in people's lifestyles, there has been a significant interest in the production of fast food and snack foods. Customers often search for simple-to-prepare foods, such as snacks, which are defined as quick meal options with or without substantial nutritional content. Making novel snack bars with beneficial components has gained popularity in recent years. Snack bars can therefore be considered a functional product category as well as an affordable and suitable ready-to-eat product for consumers. Snack bars can be consumed as a meal parts (as a part of breakfast, lunch, or dinner or as a snack between meals), as a dessert (after lunch or dinner), or as a meal replacement (for breakfast, lunch, or dinner). The age, gender, and nutritional knowledge of the consumer are common factors that influence snack bar consumption. The bars are often produced with a base of proteins (milk dairy proteins, soy, or whey) or cereals (oats, rice, maize) and then fortified with vitamins, minerals, and other nutrient- or energy-rich components. Snack bars are preferred not only for their portability but also for the possible adverse impacts on health when consumed. Additionally, nutritional snacks can offer energy and micronutrients to both healthy people and people living in famine-affected regions of the world.

Different formulations for snack bars are shown in the table:

Snack bar type	Formulation and Functionality
Fruit-based snack bar	**Ingredients:** Base: rice crisps, glucose syrup, honey, vegetable oil, quick-cook rolled oats (or natural apple dietary fibre or inulin), glycerol, whey protein concentrate, malt dextrin, pectin; Filling: pectin, sugar, citric acid, glucose syrup, honey, vegetable shortening, glycerol, and apple puree. **Process:** Base: the dry and wet ingredients are mixed, moulded and baked at 130°C/15 min; Filling: ingredients are heated with stirring to 84-86% soluble solids content; Assembly: the filing is placed between base parts **Functionality:** Increased polyphenol and dietary fibre contents
Wheat- or soy-based bar	**Ingredients:** water, corn syrup, glycerin, brown sugar, arabic gum, emulsifier, puffed wheat, wheat germ, commercial coatings, shortening, soy nuggets, soy protein, soy fibre, vanilla extract **Process:** ingredients (water, corn syrup, glycerin, brown sugar, Arabic gum, emulsifier, shortening, vanilla extract) are mixed and blended in different proportions to obtain wheat bar; The puffed wheat, wheat germ, soy nuggets, soy protein, soy fibre, are heated at 85°C/4 min. All the

	ingredients are mixed together and moulded. The bars are cooled at room temperature. **Functionality:** High quality proteins, fibres and B-complex vitamins contents
Cereal snack bar	**Ingredients**: corn starch biscuit, skimmed milk powder, rice flakes, oat flakes, corn syrup **Process**: dry ingredients are mixed and then syrup is added. The bars are manually sealed, cooled for 8 h and then cut and packed in aluminium-coated cellophane **Functionality:** Cereal flour provided an increase in dietary fibre content, vitamin C, minerals, and antioxidant activity.
Vegetable based snack bar	**Ingredients:** Bean flour, oat flakes powder, water and soy lecithin **Process:** The ingredients are mixed, and the dough obtained is coated and cut to a standard size. The snacks are baked using convection oven at 120°C/30 min, cooled at room temperature and packed in metallic bags **Functionality:** Increasing in protein and dietary fiber content and the antioxidant capacity
High-protein snack bar	**Ingredients:** Milk protein, high fructose corn syrup and glycerol. **Process:** the ingredients are mixed, and the mixture obtained is moulded. **Functionality:** Increasing in protein content

Cheese

Cheese is dairy product formed by coagulation of casein, milk protein. It is available in wide range of flavors and textures; it mainly has fat and proteins from the milk. Cheese has more value because of its longer shelf life, high protein and fat content, calcium and phosphorous. The fermentation of milk into cheese is mostly carried out by certain lactic acid bacteria, whose main function is to turn lactose into lactic acid, which lowers the pH of the curd. The pH of the finished product varies from 4.6 to 5.3 depending on the curd's ability as a buffer.

One of the most important aspects in the cheese-making process is selecting the right kind of milk. While buffalo milk can also be used if specific process modifications are made, cow milk is typically used worldwide for producing cheese. The quality of the raw milk used for producing the cheese is what determines its quality. The number of processes, including rennet coagulation, starter growth, and texture changes throughout cheese ripening are affected by the quality of the milk. Organoleptic, microbiological and chemical tests are used to evaluate milk quality.

Types of cheese

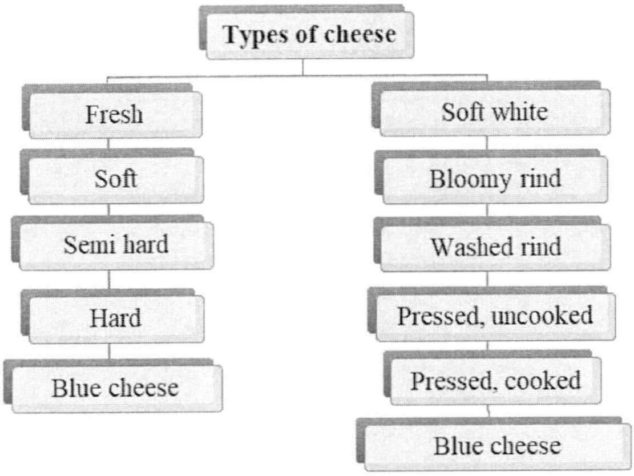

Fig 1: Type of cheese

a) **Soft white:** It has not yet matured or fermented. It is creamy and smooth, based on cow's milk. Skimmed or non-skimmed, sometimes enriched with cream.

b) **Bloomy rind:** It is not cooked or pressed and the tastes is salty.

c) **Washed rind:** During ripening, washed-rind (rind is an outside layer formed during aging) cheeses' surfaces takes on an orange to crimson colour that is caused by bacteria and yeasts. This kind of cheese is produced from raw or pasteurized milk that originates from the northern region of France. It matures for 2 to 6 months before being washed in brine that has a subtle salt flavor. Spicy in taste and aroma.

d) **Pressed, uncooked:** The raw milk is heated to a temperature of 36^0 C and coagulated at a slightly lower temperature. During the first two to three months of maturation, the rind is rubbed to create a uniform patina and reduce humidity.

e) **Pressed, cooked:** It is a hard-pressed, cooked cheese that is produced using evening milk that has been skimmed and let to stand overnight followed by mixing with the next morning milk. The maturation takes place in a cool humid cellar and lasts for Four to ten months, which involves scraping and washing the cheese in brine with a low salt content.

f) **Blue cheese:** Blue mould is a strain of penicillium that is added to the milk before rennet is added, either in liquid or powder form. The majority of blue cheese is wrapped in foil to prevent drying out and is not pressed nor cooked.

g) **Natural rind cheese:** The rind of goat's and sheep's cheese is somewhat wrinkled and has a cream colour. As the cheese dries, the wrinkles become more apparent and the character and flavour develop. Taste is fresh, almost fruity.

h) **Processed cheese:** An unfermented dairy product, such as emulsifiers, additional salt, food colouring and whey, is added to regular cheese to produce processed cheese, also known as process cheese or prepared cheese. Processed cheese comes in a variety of tastes, colours, and textures.

Generalized cheese manufacturing procedure

The temperatures, times, and target pH for different steps, the sequence of processing steps, the use of salting or brining, block formation, and ageing vary significantly between cheese types. Three basic phases make up the production of cheese. The first is casein precipitation into curd. Warm milk is infused with lactic acid-producing bacteria to increase the amount of acidity needed for rennet activity. Then, rennet is added, which causes the caseins to aggregate, allow trapping whey and fat globules in the protein network. The concentration of curds is the second step. Whey that isn't used is drained out. Cut, squeezed, cooked, and salted. The green curd's ripening or ageing is the last stage. It turns the bland, crumbly, or rubbery curds produced initially into a smooth material with a strong, complex flavour. The majority of the molecular breakdown that occurs during ripening is brought on by the enzymes of microbes, including both the initial starter bacteria and peculiar ripening organisms.

The process for making all forms of cheese is mostly the same (Figure 2), with a few steps modified to generate an end product with the desired properties. The following is a description of a few important steps in the production of cheese:

1) **Preparation of milk:** Cheese is a high protein dairy food that is produced mostly from milk from animals such as cows, sheep, goats, buffalo, and yaks, among others. Milk must be prepared before being manufactured; this is done by pasteurizing, homogenizing, and finally clarifying the milk.

2) **Adding a starter:**

Two techniques are used for this

- Sour milk gets thicken with the addition of lactic acid bacteria.
- The majority of cheeses are of the sweet milk variety, popularly known as "rennet cheeses," because they are made by adding

rennet, an enzyme obtained from the stomach of nursing calves, to separate the milk's solids from the liquid. Rennet causes milk to curdle without going sour. It also causes the milk's protein to buildup. Today, a calf's gut lining provides the majority of the rennet used in cheese production.

3) **Formation of coagulum:** When starter is added, milk coagulates into a thick mass known as "young curd," and whey separates. When the procedure is finished, the liquid whey is drained off, and cheese is made from the solid curd or coagulated milk protein mass. This is often referred as "firm curd" at this point.

4) **Cutting:** With the help of knives, firm curd is divided into smaller pieces. The type and recipe of the cheese will determine the cut's size. The majority of the time, this process is done by hand. Mainly for hard cheeses, this procedure is used.

5) **Stirring/Scalding:** For hard cheeses, this procedure is used. The curd shrinks and it releases more whey. The metabolism of the bacteria is accelerated by this procedure.

6) **Salting:** Salt is added into the cheese by wet or dry method as per the recipe. Brining in some cheeses also leads to longevity in shelf life. The cheese is salted by either sprinkling coarse salt on top of it or immersing pressed cheese in a salt solution. In addition to completing the cheese's drainage process, salting has an impact on enzyme activity, rind development, and the growth of bacteria.

7) **Moulding or Vatting and Pressing:** Cheese is placed in moulds after being salted so that it acquires a specific form. This can be carried out in wooden or plastic moulds. Pressing of cheese is done to give a distinct shape. Blue chesses are not pressed.

8) **Finishing:** Cheese is removed from its mould and given a rind or covering. The cheese may be covered, wrapped, or bandaged during this process. As they mature, the softer cheeses obtain a natural rind. Sometimes, grape must, rubber ash, and leaf wrapping are used to dry the rind. For e.g. Gorgonzola, which has a Plaster of Paris coating.

9) **Ripening/Maturing:** The majority of the harder cheeses are aged in caves. A hard cheese might take 8 weeks to a year to ripen and mature.

10) **Holding and storage:** People typically store their cheese on a stone under a clay cover in the coldest section of the house. Cheeses are held and kept in a refrigerated environment. Their tendency to dry up

and harden when exposed to air is their main drawback. It means that as soon as a block of cheese is cut into, you must either utilize it right away or carefully cover it in plastic wrap. Cheeses that are hard and firm can be refrigerated and kept for a week to many months if they are wrapped. Even with the greatest care, a whole cheese tends to dry out after being cut, acquire a sharper flavour, and form inedible moulds on the edges. Fresh cheese shouldn't be kept for longer than a week; semi-soft and soft cheeses last for around a week to two weeks.

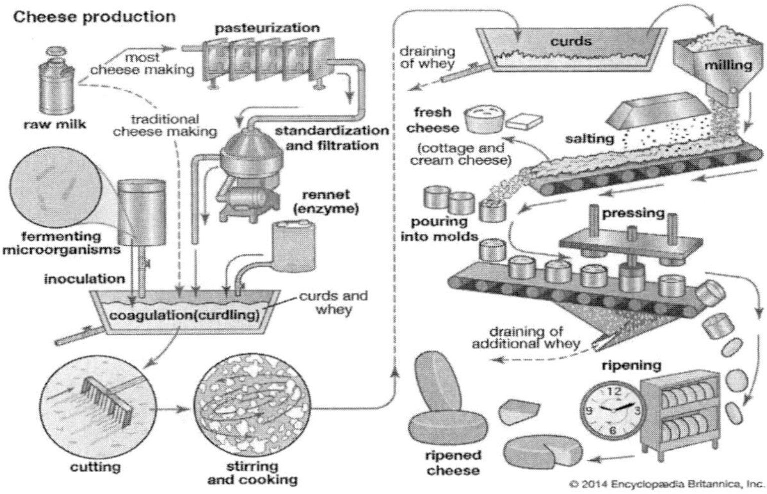

Fig 2: The process for making cheese

Conclusion

Food preservation has always been essential to our culture. Foods like jams, jellies, and preserves have a variety of textures, tastes, and colours. They are primarily made of thickened or jellied fruits that have been substantially preserved with sugar. Making novel snack bars with healthy ingredients has become more common in recent years. There are a plenty of high protein, high energy granola and cereal bars in the market. In addition to satisfying rising consumer demand, this will address a number of nutritional inadequacies that are present in developing countries. This makes the development of cereal bar formulation a new force in this specialized industry.

References

1. Biswas A. (2021). A Review paper on Fruit Jam products. International Journal of Science and Research, 7(5), 2395-4396.

2. Durrani AM, S.K., & Verma S. (2011). Development and quality evaluation of honey based carrot candy. Journal of Food Science and Technology, 48(4), 502-505.

3. Kumar A, & Deen B, (2006). Studies on preparation and storage of jelly from wood apple (*Limonia acidissima* L.) fruits. Journal of Pharmacognosy and Phytochemistry, 6(6), 224-229.

4. Kumar H., R.S, & Goraya R.K., (2017). Development and quality evaluation of guava (cv. Punjab Pink) candy. International Journal of Pure and Applied Bioscience. 5 (6), 119-126.

5. Mondhe D, S. A., Bhandare A., & Tambe M. S., (2018). Development & Quality Evaluation of Jelly Prepared from Guava Blended with Pomegranate. Iconic Research and Engineering Journals, 2(6), 322-343.

6. Srivatsava R.P. & Kumar S. (2006). Fruits and vegetables Drying/ dehydration and concentration. Fruit and Vegetables Preservation, Principles and Practices. 3rd Ed. IBDC publishers, Lucknow, UP, India. pp. 127-158.

7. Tiwari, A., Afroz, S. B., & Kumar, V. (2021). Market vulnerabilities and potential of horticulture crops in India: With special reference to top crops. Indian Journal of Agricultural Marketing, 35(3), 1-20.

8. https://projectreports.eiriindia.org/product/technology-confectionery-chocolates-toffee-candy-chewing-bubble-gum-lollipop-jelly-products-formulations-e-book/

9. http://www.eagri.org/eagri50/HORT381/pdf/lec15.pdf

Chapter - 3

Preservation by Sugar- Importance- Principles of Gel Formation- Preparation of Jam, Jelly, Marmalades, Preserve, Candy, Glazed and Crystallized Fruits

Authors

Shruti Mallikarjun Kolur

Ph.D. Scholar, Department of Horticulture, University of Agricultural Sciences, GKVK, Bangalore, Karnataka, India

Ashwini S.

Ph.D. Scholar, Department of Plant Pathology University of Agricultural Sciences, Dharwad, Karnataka, India

Santosh M Kolur

MSc. Scholar, Department of Plant biotechnology, University of Agricultural Sciences, GKVK, Bangalore, Karnataka, India

Chapter - 3

Preservation by Sugar- Importance- Principles of Gel Formation- Preparation of Jam, Jelly, Marmalades, Preserve, Candy, Glazed and Crystallized Fruits

Shruti Mallikarjun Kolur, Ashwini S. and Santosh M Kolur

Abstract

Sugar helps in the preservation of products in which it is used. The high osmotic pressure of sugar creates conditions that are unfavorable for the growth and reproduction of most of the species of bacteria, yeast and moulds. When sugar absorbs most of the available water from microbial cells or prevents normal diffusion of water into these cells with the result that there is very little water for the growth of microorganisms. Hence their multiplication is inhibited, and even though already present dies out gradually. Usually, 70% sucrose solution will stop growth of all microorganisms in foods. Thus, sugar acts as a preservative by osmosis and not as a true poison for microorganisms. Fruit syrup, Jam, Marmalade, Preserve, candy, crystallized fruit and glazed fruit are preserved by sugar.

Keywords: Sugar, Fruit syrup, Preservation, Micro-organisms

1. Introduction

Any method that involves prevention of food from spoiling due to microbial or enzymatic activity, as well as other physical activity, is referred to as food preservation. There are several techniques that helps to preserve food from spoilage including pasteurization, curing, canning, smoking, pickling, irradiation, smoking, addition of sugar *etc.* Sugar contributes distinctive sweetness to food products and enhances deliciousness. It provides a source of energy for nitrate converting bacteria.it lowers the acidity of the food.

Osmosis causes sugar in high amounts to serve as a preservative. Water is moved from the bacteria into the concentrated sugar syrup because sugar draws all available water. The microflora is dried out and unable to grow. In order to prevent the growth of microbes, sugar preserved products must have a sugar concentration of 68% or higher. Lower quantities might work, but only temporarily, unless the food contains acid or is refrigerated. The type of microorganisms and the presence of additional food components affect the critical sugar concentration needed to limit microbial development.

Some of the most popular preserves with sugar are jelly, jam and marmalade. These are the stable gels. Pectin, a natural component of fruits, forms a gel only in the presence of sugar and acid. Sugar prevents spoilage of jams, jellies and preserves even after the container is opened. Most of these products are made of acid fruits. When foods low in acid are used, they are usually combined with some acid fruit. Besides contributing flavour to the product, the acid aids in the preservation. The amount of sugar used in manufacture of these products varies widely.

Sugar may be included in the form of

a) Sucrose (table sugar/brown sugar).

b) Dextrose (refined corn sugar).

c) Corn syrup solid

Different fruits and vegetables preserved by sugar are:

- Fruits that are frequently preserved by the sugaring method include apple, pears, cherries, plums, apricots, and peaches. Before preservation, these fruits are either dried or sugar syrup-glazed.
- Using fruits jams, jellies, marmalade or wines also prepared with the help of sugar.
- Lemon peels are frequently candied to make relishes. However, entire amla is included in the Indian candied dish "murabba".
- The most frequently sugared vegetables are carrot and ginger, which are used to make sweet pickles and relish.

2. Advantages of sugar preservation

- It is also a quick and simple preservation technique.
- Since sugaring uses fewer components than other preservation techniques and frequently uses sugar extract or glaze to sweeten a variety of other foods, it has advantages over other preservation techniques.

3. Disadvantages of sugar preservation

- There is a risk in sugar preservation method as sugar is believed to attract moisture very fast.

- When the amount of atmospheric moisture is high, the environment's yeast begins to work, and sugar begins to ferment into alcohol and carbon dioxide.
- Although fermented food is also a preserved food, the sugared foods should be prevented from fermenting, as it may lead to an unpleasant taste.

4. Principle of gel formation

Gel is an intermediate substance between solid and liquid phase. It possesses both elastic (solid) and flow (liquid) properties. When gels are created, a sol (liquid) to gel (solid) transition occurs. Many foods are marketed to consumers in the form of gels because it is more convenient. Jam, jelly, confectionary items, sweets, quick-set gels and products made from fruit are some examples. Gel formation is produced by the relationship between pectin, water in the fruit, and sugar, under a controlled pH. High methoxyl pectins gel at acid pH (less than 3.5) in the presence of sugar. The desired or targeted parameter(s) are always achieved using one or more gelling agents. These goods generally contain gelling ingredients, which are typically allowed food hydrocolloids. These hydrocolloids are derived from natural sources and refers to a range of polysaccharides and proteins. Examples of gelling agents: Gelatin, Whey protein, Egg proteins, Zein, Soya proteins, Agar, Cereal flour, starch, *etc.*

4.1 Mechanism of gel formation

A food gel can also be considered as a high moisture three dimensional polymeric network that resists flow under pressure and largely maintains its distinctive structural shape (mechanical stiffness). The most commonly accepted theory of gel formation is that the mechanism of stacking of polysaccharide chains to form junction zones. For gel formation sugar, acid, water, pectin must be present. Protons of the acid shift the equilibrium between ionized and unionized groups. Added sugar further decreases the hydration of pectin by competing for water. Hence water is less free to solvate the polysaccharide and so there is increased hydrophobic interaction between methyl esters group. As a result of losing part of their charges and being more hydrated, the polymer molecules can now join along some of their length, establishing junctions, and a network of polymer chains is created that traps solute molecules in an aqueous solution.

4.2 Role of physico-chemical factors in gel formation

a) **Temperature:** Obtaining gels through heat-induced gelation is likely the most important and popular way. A two-step process, gelation begins with the molecules unfolding or dissociating as a result of the energy input, exposing the reactive sites. The interaction and aggregation of unfolded molecules to create complexes with larger molecular weight constitutes the second phase.

b) **Pressure:** Since high pressure can be used alone or in combination with other processes, particularly with higher temperatures, it provides an additional degree of flexibility when changing the functional properties of molecules. High pressure generally encourages reactions, which reduce the system's overall volume. Water dissociates under pressure, and the pH increases under pressure. Different gels created by applying pressure and heat have different rheological characteristics (Heremans, 1995).

c) **Ionic strength:** Calcium and sodium are two monovalent and divalent cations that can strengthen the gel's ionic bonds. Gelation is possible when the electrostatic repulsive interactions between the molecules are diminished or neutralized (Alting *et al.*, 2002).

d) **pH:** The net charge of a molecule is altered when the pH changes as a result of the addition of acids or microbial fermentation, which changes the attractive and repulsive forces between molecules as well as the interactions between molecules and solvents, or hydration characteristics (Lucey and Singh, 1997).

e) **Presence of enzymes:** The introduction of enzyme-induced gelation is the insertion of synthetic covalent cross-links into dietary proteins. Protein cross-linking can be achieved, among other things, through processes facilitated by trans-glutaminase (TG), peroxidase, and polyphenol oxidase (Lauber *et al.*, 2003).

f) **Solvent quality:** Gel formation is significantly influenced by the type and amount of solvent present. For instance, concentrated sugar solution can create gels at concentrations as low as 0.2%, whereas acid-thinned starch needs a concentration of about 15% to form gel.

5. Jam

It is essentially a gel or semi-solid mass made from crushed or macerated (ground) fruit and generally is less firm than jelly. They may be made from a single fruit or mixture of fruits. Jam is made by cooking fruit pulp with enough sugar to achieve a relatively thick consistency that is solid enough to keep the fruit tissues in place. The FPO specification of jam is 68.5% TSS, 45% of fruit pulp and 0.5-0.6% of acid (citric acid) per 100gm of the prepared product. Apple, sapota, papaya, plums, mango, grapes, jack, pineapple, banana, guava and pears are commonly used for preparation of jam.

Procedure for preparation of fruit jam

1) **Selection of fruits:** Select fully ripe fruits having good colour, flavour and aroma.
2) **Preparation of the fruit:** In clean water, properly wash the chosen fruits. Cut off the swollen and discoloured areas. Fruits should be peeled, with any stones and cores removed. Fruits should be cut into small pieces. Boil the fruit in a tiny amount of water to soften it if it is harsh and hard. Put the fruit pieces in a blender and process until they are a fine pulp.
3) **Addition of sugar and acid:** To the sour fruits add 1kg of sugar to each kg of pulp, while to the sweeter fruits add only 3/4th kg sugar to each kg of pulp. Add citric acid and sugar as given in the recipe. Mix the ingredients properly.
4) **Cooking:** Cook the mixture slowly with occasional stirring and crushing, till the temperature reaches 105.5^0 C at mean sea level or till the mass reaches the desired consistency. The temperature is adjusted at different places by deducting 0.4^0 C for every 100 m rise in the altitude from the sea level.
5) **End point:**
   i) **Temperature test:** The bulk should be heated until it achieves the necessary temperature for that height. Boiling can be stopped once this temperature is achieved.
   ii) **Sheet test:** A small portion of jam is taken out during boiling, in a spoon or wooden ladle, and cooled slightly. It is then allowed to drop. If the product falls off in the form of a sheet or flakes instead of flowing in a continuous stream or syrup, it means that the end-point has been reached and the product is ready, otherwise, boiling is continued till the sheet test is positive.
6) **Filling and closing:** Fill the hot jam into sterilized, clean, dry jam bottle placed on an insulating material like wooden board or a container without any delay. Cool the filled bottle in air before storing.
7) **Labeling and storage:** Label the bottles according to specifications like kind of product, date of manufacture, net weight of the product and chemical preservatives used if any. Store them in a cool dry place.

5.1 Flow chart of jam preparation

Ripe firm fruits → Washing → Peeling →Pulping (Remove seed and core) → Addition of sugar and acid → Boiling (with continuous stirring) → Judging of end point by further cooking upto 105o C (or) 68% TSS (or) by sheet test → Filling hot into sterilized bottles → cooling → Sterilized bootles → cooling → Capping → Storage (at ambient temperature).

5.2 Problems in Jam Production

1) **Crystallization:** The final product should contain 30-50% invert sugar. If it is less than 30% cane sugar may crystallize out on storage and if it is more than 50% the jam will become a honey like mass due to the formation of small crystals of glucose. Corn syrup or glucose may be added along with cane sugar to avoid crystallization.

2) **Sticky or Gummy jam:** Because of high percentage of total soluble solids, jams tend to become gummy or sticky. This can be solved by addition of pectin or citric acid or both.

3) **Premature setting:** This is due to low TSS and high pectin content in the jam and can be prevented by adding more sugar. It is this cannot be done a small quality of sodium bicarbonate is added to reduce the acidity and thus prevent pre coagulation.

4) **Surface graining or shrinkage:** This is caused by evaporation of moisture during storage of jam. Storing in a cool place can reduce it.

6. Jelly

A jelly is a semi-solid product prepared by boiling a clear, strained solution of pectin containing fruit extract, free from pulp, after the addition of sugar and acid. FPO specification for jelly is Acid content should be 0.5 – 0.75%, TSS 65% and fruit juice 45%, 0.5 – 1.0% pectin.

Fruits suitable for jelly preparation: Guava, sour apple, plum, karonda, wood apple, loquat, papaya and gooseberry are normally used for preparation of jelly. Apricot, pineapple, strawberry, raspberry etc. can be used but only after addition of pectin powder, because these fruits have low pectin content.

6.1 Characteristics of jelly

- A perfect jelly should be transparent well set but not too stiff.
- The fruit's natural flavour should be present.
- It should be firm enough to retain a sharp edge and tender enough to quiver when pressed.

- The product should have no syneresis (weeping) and neither tough.
- When taken from the moulds, it should maintain its shape and have a pleasing colour.
- It shouldn't be syrupy, gummy, sticky, or include crystallised sugar.

6.2 Flow chart for jelly preparation

Fruit (Firm, not over ripe) → Washing → Cutting into thin slices → Boiling with water (1 ½ times the weight of fruits for about 20-30 min) → Addition of citric acid during boiling (2 g per kg of fruit) → Straining of extract → Pectin test (for addition of sugar) → Addition of sugar → Boiling → Judging of end point (sheet / drop / temp test) → Removal of scum (or) foam → Addition of colour and remaining citric acid → Filling hot into clean sterilized bottles → Capping → Storage at ambient temperature.

6.3 Problems in jelly making

- **Formation of crystals:** It is mainly due to both the excessive sugar addition and the over concentration of jelly.
- **Syneresis or weeping of jelly:** Syneresis, often known as weeping, is the term used to describe the phenomena of a jelly's fluid exuding on its own. It is caused due to excess of acid, too low concentration of sugar, insufficient pectin and premature gelation.
- **Cloudy or foggy jellies:** It is mainly because of over cooking, non-removal of scum, over cooling, use of non-clarified juice or extract and use of immature fruits.

7. Marmalade

This is a fruit jelly in which slices of the fruit or its peel are suspended. The term is generally used for products made from citrus fruits like oranges and lemon in which shredded peel is used as the suspended material. FPO specification for marmalade is 65% TSS, 45% juice or extract.

Types of marmalade:

1) **Jelly marmalade:** In jelly marmalade only extract is used
2) **Jam marmalade:** In jam marmalade, pectin extract of fruit is not clarified and the whole pulp is used.

7.1 Flow sheet for marmalade preparation

Selection of Ripe fruits → Washing → Peeling outer yellow portion → Cutting yellow → portion into fine shreds → Cutting of 0.3-0.45cm thick slices of peeled fruit or crushing into pulp → Boiling Straining the extract →

Testing for pectin content→ Addition of sugar→ Cooking to 103 to 105 ^{0} C→ Addition of prepared shreds→ Boiling till jellying point→ Testing for end point→ Cooling → Flavouring→ Filling hot into clean sterilized bottles → Capping → Storage at ambient temperature.

7.2 Problems In Marmalade preparation

Browning: It occurs frequently when being stored and may be avoided by adding 0.09g of KMS per kilogramme of marmalade and avoiding using tin containers. KMS is incorporated into the marmalade as it cools after being dissolved in a tiny amount of water. Additionally, KMS removes the potential for mold-related deterioration.

8. Preserve

A mature fruit or vegetable or its pieces impregnated in heavy sugar syrup till it becomes tender and transparent is known as preserve. Aonla, bael, apple, pear, mango, cherry, karonda, strawberry, pineapple, papaya etc are used for preserve making. FPO specification for preserve is 68% TSS, 55% fruit juice or pulp.

Fruit cooked in syrup is challenging because the syrup needs to be kept at the right consistency to saturate the entire fruit without shrinking or toughening it. Fruit shrinks when cooked in syrup directly, which also limits the sugar's absorption. Therefore, before steeping in syrup, the fruit needs to be blanched to make it soft enough to absorb water. Fruits with a lot of juice, however, can be cooked straight.

8.1 Fruits may be cooked in syrup by three processes

A. Rapid process

Fruits are boiled in syrup with little sugar. The syrup is allowed to boil with gentle heating until it thickens enough. It is safe to cook soft fruits like strawberries and raspberries in heavy syrup since they may be softened more quickly than hard fruits like apples, pears, and peaches, which need to be heated for a long time. However, rapid boiling should be avoided because it makes the fruit harsh, particularly when heating is done in a big shallow pan with little syrup. A final sugar concentration of not less than 68%, or a boiling point of 106 ^{0} C, is required. Although this is a quick and inexpensive method, boiling significantly reduces the product's flavour and colour.

B. Slow process

The fruit is blanched until it becomes tender. After that, alternate layers of sugar equal to the weight of the fruit are added, and the mixture is let to stand for 24 hours. During this time, the fruit releases water and the sugar dissolves, creating a syrup with a TSS content of 37–38%. The syrup is heated the following day after the fruits are removed to increase its strength to about 60% TSS. To invert some of the cane sugar and avoid crystallisation, a tiny amount of citric acid (1 to 1.5g per kilogramme sugar) is also added. Following a 4-5 minute boil, the entire mass is then refrigerated for 24 hours. On the third day, the syrup's potency is increased.

C. Vacuum process

The fruit is first softened by boiling and then placed in the syrup which should have 30-35% TSS. The fruit-syrup blend is then transferred to a vacuum pan and concentrated under reduced pressure to 70% TSS. Preserves made by this process retain the flavor and colour of the fruit better than by the other two methods.

In all of these procedures, the fruit is covered with syrup both during and after cooking. It will eventually dry out, which will lower the product's quality. The product should be cooled quickly after the final boiling to prevent discolouration during storage. The fruits are drained free of syrup and filled in dry containers or glass jars. Freshly prepared boiling syrup containing 68% TSS is then poured into the jars/ containers which are then sealed airtight.

8.2 Flow chart of preserve preparation

Mature fruits → Washing → Preparation of fruit for sugar treatment → Keeping fruit and sugar in alternate layers (1.0 kg Fruit: 1 kg Sugar) (or) steeping fruit in syrup of 40% TSS for a day → Removal of fruit → Increasing consistency of syrup to 60% TSS by boiling → Steeping of fruit for a day →Repeating the process and raising strength of syrup by 5% TSS to 70% on alternate days – Steeping in 70% TSS for a week → Preserve – Draining – Filling in jar (or) container → Covering fruit with freshly prepared sugar syrup of 68% TSS →Sealing (airtight) – Storage.

9. Candy

A fruit or vegetable impregnated with cane sugar or glucose syrup, and subsequently drained free of syrup and dried is known as candied fruit or vegetable. Aonla, karonda, pineapple, cherry, papaya, apple, peach and peels of orange, lemon, grape fruit and citron, ginger *etc* are used for candy preparation.

FPO specification for candy is 70% TSS, 25% juice or pulp.

Preparation: Preparing candied fruit is basically identical to preparing preserves. The fruit impregnated with syrup has a higher percentage of sugar

(75o bx), which is the only difference. After an appropriate dilution, the syrup leftover from the candying process can be used to make vinegar, sweeten chutneys, sauces, and pickles, and candy another batch of the same type of fruit.

Examples of candied fruit/vegetable products: Candied citrus peels, Petha (Candied ash gourd) *etc.*

10. Glazed fruits/vegetables

Covering of candied fruits / vegetables with a thin transparent coating of sugar, which imparts them a glossy appearance, is known as glazing.

Preparation: Cane sugar and water (2:1 by weight) are boiled in a steam pan at 113-114°C and the scum is removed as it comes up. Thereafter the syrup is cooled to 93°C and rubbed with a wooden ladle on the side of the pan when granulated sugar is obtained. Dried candied fruits are passed through this granulated portion of the sugar solution, one by one, by means of a fork, and then placed on trays in a warm dry room. They may also be dried in a drier at 49°C for 2-3 hours. When they become crisp, they are packed in airtight containers for storage.

11. Crystallized fruits/ vegetables

Candied fruits/vegetables when covered or coated with crystals of sugar, either by rolling in finely powdered sugar or by allowing sugar crystals to deposit on them from a dense syrup are called crystallized fruits.

Preparation: The candied fruits are placed on a wire mesh tray which is placed in a deep vessel. Cooled syrup (70 per cent total soluble solids) is gently poured over the fruit so as to cover it entirely. The whole mass is left undisturbed for 12 to 18 hours during which a thin coating of crystallized sugar is formed. The tray is then taken out carefully from the vessel and the surplus syrup drained off. The fruits are then placed in a single layer on wire mesh trays and dried at room temperature or at about 49°C in driers.

Conclusion

Sugaring is an old-fashioned curing or fermenting process for preserving fruit and vegetables. Early in the 20th century, notably during the First and Second World Wars, this method of food preservation was highly common. Sugar also provide an indirect form of preservation by serving to accelerate accumulation of antimicrobial compounds from the growth of certain other microbial organisms. Examples include the conversion of sugar to ethanol in wine by fermentative yeasts or the conversion of sugar to organic acids in sauerkraut by lactic acid bacteria by the process fermentation. Sugar itself doesn't preserve food it undergoes the process called as osmosis which causes dehydration of food help in preserving food.

References

1. Heremans K. (1995). High- pressure effects on biomolecules. *In: High Pressure Processing of Foods,* pp. 81–98.
2. Alting A.C, De Jongh H.H.J, Visschers R.W. and Simons J.W.F.A. 2002. Physical and chemical interactions in cold gelation of food proteins. *Journal of Agricultural and Food Chemistry,* 50: 4682-4689.
3. Lucey J.A. and Singh H. (1997). Formation and physical properties of acid milk gels: A review. *Journal of International food research,* 30: 529-542.
4. Lauber, S., Krause, I., Klostermeyer, H. and Henle, T. (2003). Microbial transglutaminase cross links β-casein and β-lactoglobulin to heterologous oligomers under high pressure. *European Food Research and Technology*, 216: 15–17.

Chapter - 4

Unfermented fruit Beverages: Preparation and Preservation, Squash, RTS Beverages, Nectar, Cordial, Syrup, Crush, Concentrate. Methods of Extraction, Clarification and Pasteurization

Authors

Raju C.A

Ph.D. Scholar, Department of Food Science and Nutrition, University of Agricultural Sciences, GKVK, Bangalore, Karnataka, India

Ashoka S

Ph.D. Scholar, Department of Food Science and Nutrition, University of Agricultural Sciences, GKVK, Bangalore, Karnataka, India

Ganavi B.R

Ph.D. Scholar, Department of Food Science and Nutrition, University of Agricultural Sciences, GKVK, Bangalore, Karnataka, India

Pavuluri Yasaswini

Ph.D. Scholar, Department of Food Science and Nutrition, University of Agricultural Sciences, GKVK, Bangalore, Karnataka, India

Chapter - 4

Unfermented fruit Beverages: Preparation and Preservation, Squash, RTS Beverages, Nectar, Cordial, Syrup, Crush, Concentrate. Methods of Extraction, Clarification and Pasteurization

Raju C.A, Ashoka S, Ganavi B.R and Pavuluri Yasaswini

Fruit drinks have become popular over carbonated and alcoholic beverages by consumers because they are regarded to be healthier. As a result, the fruit juice category is anticipated to see rising demand in terms of both volume and value, as the total market return is mostly from Asian and Latin American nations. With post-harvest losses typically ranging from 25–30%, India is the nation with the second-greatest output of fruits and vegetables in the entirety of the globe, after China. In India, merchandise with added value is created using less than 3% of the country's entire fruit and vegetable abundance. In light of escalating modernization and an upsurge in female employment in the workforce, the geographic reach of the field of fruit processing has broadened substantially. The most crucial element of managing surplus throughout the fruit season's harvesting process is the processing of surplus fruits into semi-finished merchandise like pulps and juices. These intermediate goods (raw puree and juices) are capable of being transformed into completed goods comprising squashes, ready-to-serve beverages (RTS), nectar, cordial, etc. during the off-season. In this sense, production and storage are important primordial pillars of the food economy. Consequently, there is an immense amount of opportunity for processing fruits into diverse products to cut down on waste and earn revenue worldwide.

Keywords: Beverages, Fruits, Nutrients, Processing, Preservatives

Introduction

Fruit is a tree or plant's edible and more or less juicy produce that comprises the developed ovary, containing its seeds and nearby components. Fruits are typically sweet, with an assortment of flavours, colours, and textures. Fruits are a fantastic and indispensable source of minerals, vitamins, carbohydrates, and phytochemicals, all of which have proven therapeutic and protective properties. Despite this, a vast majority of fruits are not long-lasting and were cultivated in geographically distant places; for example, apple, pineapple, mango, citrus, guava, and berries are grown in selected temperature zones. Fruits' nutritional qualities are not only beneficial to the regular functioning of the human body, but they additionally mitigate the risk of cardiovascular disease, control blood pressure, and downplay oxidative stress. Conversely, fruits high in dietary fibre control digestive system function and minimize the dangers of obesity, particular kinds of cancer, and visual ability loss.

Beverages are an integral part of human nutrition, even for new born babies. Infant formulas-highly complex fluids packed with many essential nutrients are where the cycle originates. Product designers adapt to the changing nutritional demands of people as they age and make new, creative beverages to fulfil those desires. The definition of a beverage is "any fluid consumed by drinking." It comprises a wide range of foods, mostly liquids, starting with the most basic beverage, "water," and a variety of commercially accessible liquids including milk, dairy products, tea, coffee, chocolate drinks, and fruit beverages. Although various beverages possess distinct qualities, they all have the potential to quench thirst. Beverages are described as "liquid that is primarily created or designed for human consumption". Despite being essential to life, beverages are rarely indulged in for their nutritional content. Although fulfilling human needs is their primary purpose, they are inevitably a part of our culture.

Fruit-based drinks and juices often include naturally extracted juices, drinks, ready-to-serve (RTS) beverages, nectars, squash, cordials, and appetisers, among other things. In comparison to many artificial and aerated drinks, these items are much more hydrating, thirst-quenching, tasty, and nutritionally better. Fruit juice is the naturally occurring liquid that the edible component of the fruit presses or otherwise extracts mechanically. Depending on their structure and content, fruits may typically be juiced in a variety of techniques.

In the sweltering summer, tropical nations like India may provide a wide variety of delectable refreshing drinks, particularly fruit-based concoctions. These fruit drinks are getting more and more well-liked and are progressively taking over a portion of the market for refreshing beverages as a result of growing consumer knowledge about quality, safety, and health.

While fruit beverages such as drinks, squashes, cordial, etc. blend fruit juice or pulp, sugar, flavor, color, acid, etc. in the right proportions to get a desired flavor, the composition of juice remains unchanged during preparation and preservation. Commercial items on the market include mango squash, orange juice, pineapple juice, apple juice, guava drink, and mixed fruit drink, mango nectar, squash, etc.

Types of fruit beverages

1. Unfermented beverages

Unfermented drinks are defined as fruit juices that do not go through the alcoholic fermentation process. They are made up of fruit juice concentrate, fruit juice powder, RTS, nectar, cordial, squash, crush, and sweetened fluids. This group also includes carbonated drinks and barley waters.

Types of unfermented beverages

1) **Fruit juice:** It can be described to be natural fruit juice, which can be apple juice, or unfermented fruit juice.
2) **Fruit juice beverage:** Prior to drinking, it may or may not be diluted. Nectar, cordial, squash, and appetizer are all examples of ready-to-serve (RTS) beverages that contain fruit juice.
3) **Synthetic drinks:** The components that are used to make synthetic beverages include sugar, water, flavorings, acidulants, ccolors *etc.* These beverages don't have any fruit juice or pulp in them.
4) **Ready to serve (RTS) drink:** Fruit drinks of this category must have at least 10% fruit juice (5% juice in lime drinks) and at least 10% total soluble solids.
5) **Fruit nectar:** This particular fruit beverage is heat-processed to maintain freshness and includes at least 20% fruit juice or pulp and 15% total soluble solids. Fruit nectars cannot include more citric acid, or 1.5 per cent, that permitted by regulation.
6) **Squash:** Fruit pulp or juice is used to make fruit squash, which is then sweetened with cane sugar. The fruit squash must include at least 25% fruit juice or pulp and at least 40% TSS in accordance with Indian Food Regulations.
7) **Fruit juice cordial:** It's a sparkling, clear, sweetened fruit beverage free of pulp and other suspended impurities. It has at least 25% juice, 30% total soluble solids, and a preservative concentration of no more than 600 ppm benzoic acid or 350 ppm SO_2.
8) **Fruit appetizer:** Although it incorporates spices, sauces, and herb extract, fruit appetisers are comparable to fruit squash. In order to

make appetisers, salt, menthe essence, big cardamom, ginger, black pepper, and cumin are utilised. Spiced fruit squash is another name for them. It's extremely customary to serve appetisers with plum and apricot.

9) **Fruit crush:** Fruit crush must have a minimum concentration of 25% fruit juice or pulp, a maximum of 55% total soluble solids (TSS), and an SO_2 or benzoic acid concentration which is no more than 350 ppm.

10) **Fruit syrup:** A minimum of 25% fruit juice or pulp is used in its preparation, and cane sugar is used for sweetening.

11) **Synthetic syrup:** All synthetic syrups are made with a foundation of heavy sugar syrup that is 70-75 percent strength and are then colored and flavored with artificial colors and flavors. Fruit juice or pulp may or may not be present in them. Sherbets and syrups made from sandal, almond, rose, khuskhus, and kewra are quite popular.

12) **Carbonated fruit beverages:** It is a ready-to-drink fruit juice beverage that comprises varying proportions of fruit juice, sugar, acid, and other ingredients and is carbonated. Carbonated fruit juice drinks can be made with apple lime juice, lemon juice, juice, or grape juice. They are made either by pre-mixing or by post-mixing.

13) **Fruit juice concentrate:** It is a concentrated fruit juice obtained by removing water using evaporation, freezing, or reverse osmosis. Fruit juice concentrate may be used to make a variety of goods. In the industry, orange juice concentrate, apple juice concentrate, and so on are commercially produced.

14) **Fruit juice powder:** This is a powder produced from fruit juice. They can be processed by foam mat drying, freeze drying, or spray drying. They rehydrate easily to produce full-strength fruit juice beverages.

They are made in accordance with the Food Safety and Standards Regulation, 2011, and the Fruit Products Order (Table 1), and are described as follows:

Table 1: Fruit Products Order (FPO) standards for fruit beverages

Product	TSS (%)	Fruit juice (%)	Citric acid (%)
Unsweetened juice	Natural	100	3.5
Fruit syrup	65	25	3.5
Crush	55	25	3.5
Squash	40	25	1.5

Fruit nectar (excluding orange and pineapple)	15	20	1.5
Orange and pineapple nectars	15	40	1.5
Cordial	30	25	3.5
Sweetened juice	10	85	-
Lime/lemon Ready to serve beverage	10	5	-
Ready-to-serve beverage/drink	10	10	-
Fruit juice concentrate	32	100	-
Synthetic syrup/sherbet	65	-	-

2. Fermented beverages

Fruit juice that has undergone alcoholic fermentation by yeast includes Saccharomyces cerevisiae is a fermented fruit beverage. Ethyl alcohol has been present in the product in variable concentrations. Common fermented drinks include plum wine, grape wine, apple cider, vermouth, and others *etc.*

Extraction, clarification and pasteurization of unfermented beverages

The structure and edible section of the fruit determine the best methods for extracting juice. Thermotherapy, freezing, chilling, concentration (drying), and for some clear juices, fine filtering are some methods used for preservation. Juices can be disassembled by taking off water, bitterness, acidity, and volatile taste components, and then reassembling them to create a uniform result. The leftover fruit material after the juice has been removed can be used to make drink bases that contain fruit.

Extraction of juices

Selection of fruit: All fruits are inadmissible due to problems in extracting the juices or because the juice is of inferior quality. The flavor and keeping quality of the juice are influenced by the kind and age of the fruit, as well as the locale of production. Only completely ripe fruits are preferred. If overripe and green fruits were utilized, the juice's quality diminished.

Sorting and washing: Fruits that are diseased, damaged, or rotting are discarded or clipped. Washing with water or weak hydrochloric acid removes dirt and spray residues of arsenic, lead, and other chemicals.

Juice extraction: In most cases, the juice is recovered by crushing or shredding the fruit and pressing the crushed mixture in a basket or hydraulic press. Juice can also be extracted using a screw-type juice extractor. A fruit grater or mill, basket or hydraulic press, screw type juice extractor, rosing or burring machine, fruit pulper, and other equipment are commonly used for juice extraction. There are two kinds of extraction methods: single and double operation systems.

- **Single operation:** The prepped fruit is typically crushed and pressed in a single operation using a screw type, plunger type, or roller type press to extract the juice. Segments of citrus fruit are fed into a hopper, moved via conical screws, and the juice flows out through the perforations while the pomace exits at the end of the conical jacket. Depending on the situation, the screw-type extractor can be operated manually or with electricity. The juice that is extracted typically comprises a sizable amount of macerated pulp and is thick and murky in appearance. Citrus fruit rinds must be carefully removed in its whole to prevent the juice from becoming harsh. Using a rosing or burring machine, citrus fruits like lemon, kinnow, etc., may also be removed. To eliminate the seeds, the liquid is lastly squeezed through a heavy cloth or a sieve.
- **Double operation:** The fruits are crushed in this technique before being individually pressed. Fruit such as apples, aonlas, berries, grapes, jamuns, and phalsas, among others, are chopped or crushed in a fruit crusher, and the crushed material is then pressed using a basket press and hydraulic press.

Process factors for various fruits' juice extraction include

- Soft fruits, such as berries or tomatoes, can be pressed using a fruit press or pulped in a food processor with a juicer attachment.
- Citrus fruits are frequently reamed to obtain the clear juice.
- Harder fruits, such as pineapple, are peeled, pulped, and pressed to produce juice.
- Apple and pear fruits are smashed in a fruit grater and pressed in a hydraulic/basket press to produce clear juice.
- Passion fruit juice is made by utilising a pulper-finisher, which removes the skin and seeds from the pulp.

Fruits that are difficult to extract clear juice from, such as mango, guava, apricot, and peach, are pulped and then used to produce juice, fruit drinks, and ready-to-serve beverages. The instruments used to make fruit juices and squashes should all be rust- and acid-resistant. Copper and iron should never be used in containers because they react with fruit acids and turn products black. Utilise equipment and instruments made of stainless steel, aluminium, and other materials. Extracted juices shouldn't be exposed to the air unduly since this will ruin the colour, flavour, and fragrance and lower the vitamin content.

Deaeration: Fruit juices consist of some air, most of which is apparent on the surface of the juice and some of which is dissolved in it. A high vacuum is used to eliminate the majority of the air and other gases from the fresh juice. This is known as deaeration, and the apparatus required to do it is known as a deaerator. Because it is a costly approach, it is not being employed in India.

Clarification of juices

Fruit juices are cleared using several processes:

- **Straining or screening:** In colloidal suspensions, impure and unclarified fruit juices include varied levels of suspended material such as fragmented fruit tissue, seeds, peel, pectic debris, and proteins. The seeds and skin, which degrade the juice's flavour, are removed by sifting through a muslin screen or sieve. Fruit juice is filtered through a muslin cloth or stainless-steel mesh sieve in small-scale companies to remove coarse particles. However, power-operated screening devices or filter presses are utilized in big businesses.
- **Finishing:** To clear the citrus juices of their cloudiness, finishing is necessary. The movement of a drill moving inside a cylinder screen separates the pulp from the finished juice. Depending on the fruit's state and fragility, screen holes range in size from around 0.020 to 0.030 inches in diameter. The amount of pulp in citrus fruit juice serves as a gauge for finishing.
- **Decantation:** The clear juice is then decanted or siphoned off once the hardened juice has settled, which is the easiest cleaning procedure. Juice solids can be settled down by storing it at a low temperature for an extended length of time, producing pure juice.
- **Centrifugation:** The centrifugal force allows for the separation of the cloud particles. The light and heavy components in each layer are separated by centrifugal force as the juice containing the particles is then pumped into a basket or disc-type centrifuge. After separating the undesirable materials, the clear juice is collected.
- **Enzymes:** Freshly squeezed fruit juices include colloidal suspensions made up of proteins, pectin, starch, and plant polysaccharides. The pectinol enzyme, which converts pectin into a

soluble form that releases suspended particulates and settles down to clear the juice, is frequently employed to improve the juiciness of fruit juices as well as to purify them. Similar to this, fruit juices are processed to remove proteins and starch using amylase, a proteolytic and starch-liquefying enzyme. For acidic liquids, pectinol is more efficient. Fruit juice may be filtered in as little as an hour at temperatures between 40 and 50°C, while it takes 20 hours at 20°C.

- **Physical finishing:** Finishing agents with physical or mechanical action include kaolin, diatomaceous earth, Spanish clay, bentonite, China clay, and filter. In general, 0.5 to 0.1 percent soil is blended with fruit juice and filtered via a filter press. Ultrafiltration is a method that separates particles based on molecular weight and improves nutrient retention in juice. To improve the viscosity and to get a good juice, pectin must be enzymatically set down prior to ultra-filtration.
- **Chemical finings:** Gelatin and casein are used in fruit juice clarifying to neutralise particles that are electrically charged and to generate an insoluble precipitate with juice components. Gelatin binds tannins and casein to the acids in the juice. When used in excess, gelatin can produce cloudy juice. The gelatin solution is added and allowed to stand for 18 to 24 hours, depending on the quantity of tannin in the fruit juice, to ensure that the precipitate settles down together. The clarified juice is then extracted. Albumin (egg whites) is another possibility to clarify liquid.
- **Clarification by freezing:** Grapefruit juice includes potassium hydrogen tartrate or cream of tartar, along with pulp and skin that are removed by freezing, cooling for long-term storage, or refrigeration. Apple juice can be clarified using this procedure after it has precipitated upon freezing.
- **Clarification by heating:** Fruit juice's colloidal component coagulates while being heated and settles when it cools, making it possible to separate it using a filter press. Juice from apples and pomegranates is heated to between 80 and 85 °C for a short period of time, quickly cooled, and then filtered through a filter press to remove any impurities.

Addition of sugar

All juices are sweetened with sugar, with the exception of grape and apple. Juice can have sugar added directly or syrup can be produced by dissolving sugar in water. Cane sugar is used in the right amounts to create syrup, cordial, and fruit squash pulp or juice. Cane sugar is also used to sweeten ready-to-serve beverages (RTS) and fruit juices.

Fortification

Vitamins are sometimes added to juices, squash, syrups, and other foods to increase their nutritional content, enhance flavour, texture, or colour, and restore nutrients lost during processing. Beta-carotene (the water-soluble form) and ascorbic acid are often added at rates of 250-500 mg and 7-10 mg per litre, respectively. Beta-carotene gives orange foods an appealing orange colour, and ascorbic acid functions as an antioxidant. For a balanced flavour, certain acids are added. For all beverages, phosphoric acid is frequently utilised, whereas citric acid is frequently used for cola-style drinks.

Preservation of juice: Nectar, RTS, and fruit juices are all preserved either by pasteurization or the use of chemical preservatives. Chemical preservatives like sodium benzoate or potassium metabisulphite are used to preserve squash, crush, syrup, and cordial.

- **Pasteurization:** Fruit juices are most frequently preserved using heat. The microorganisms that taint the juice are killed or rendered inactive by pasteurization, which involves heating the juice to 100°C or just slightly below. Depending on the type of heating system, the composition of the juice, and the size of the container, fruit juices are commonly pasteurized between 75 and 88°C for durations ranging from 30 seconds to 30 minutes. Either heating at a low temperature for a long time (LTLT) or a high temperature for a short time (HTST) can be used to pasteurize food. Common techniques include:
    - **Holding pasteurization:** Using this technique, the completed juice is placed into glass bottles with adequate headspace before the containers are airtightly sealed. The bottles are pasteurized after being sealed. However, commercial-scale goods do not employ this technique.
    - **Pasteurization by overflow method:** According to this procedure, the juice is heated to a point that is approximately 2.5 °C over the pasteurisation temperature before being poured into hot, sterilised bottles. The temperature must be kept steady throughout filling and sealing. For a set amount of time, sealed bottles are treated in boiling water. The bottles are chilled after processing. The juice contracts as it cools, leaving a tiny, airless headspace in the bottle. In the food sector, this technique is employed commercially to preserve various kinds of fluids.

- **Flash pasteurization:** In this procedure, the juice is placed into containers, heated for a brief period of time at a temperature that is just 5.50°C above the pasteurisation temperature to sterilise the seal, and then chilled. The juice is heated quickly during this procedure, commonly referred to as the HTST (High Temperature and Short Time) technique. Flash pasteurisation helps to maintain juice equally clouded, lessen the ripe flavour of the juice, and improve nutrient retention.
- **Aseptic processing and packaging of fruit juice:** The process of packing a commercially sterilised product in a pre-sterilized container in a sterilised setting is known as aseptic processing and packaging. This technique uses high-temperature, short-time (HTST) sterilisation for acidic goods (pH 4.6) throughout a temperature range of 90 to 110 °C and ultra-high temperature (UHT) sterilisation at 121 °C for low acidic meals (pH > 4.6). A continuous, closed system is used for the commercial aseptic sterilisation procedure. Better nutrient retention and great sensory quality can be achieved by aseptic processing. Tetra Pack products including apple, mango, litchi, and pineapple drink are produced for the market utilising an aseptic processing and packaging technology.
- **Preservation with chemicals:** The two most prevalent chemical preservatives are as follows:
  - **Benzoic Acid (benzoates):** The sodium salt of benzoic acid, which is soluble in water but more potent than benzoic acid itself, is frequently utilised. Benzoic acid works better against yeast than it does mould. It does not, however, prevent the fermentation of lactic acid and acetic acid. The kind of microbial illness, the acidity, and the composition of the juice all affect how much sodium benzoate is needed.
  - **Sulphur dioxide (Sulphites):** Sulphur dioxide is frequently obtained from potassium meta-by-sulphite ($K_2O_2SO_2$). Along with fruit juices or beverages, it combines with the acid in the juice to produce potassium salts and sulphur dioxide, which are then liberated and combined with the juice's water to create sulphuric acid. In addition to inhibiting enzymes and other things, sulphur dioxide is more efficient than yeast at killing bacteria and mould spores. As a bleaching and antioxidant, it aids in retaining ascorbic acid, carotene, and other oxidizable

substances. The product is additionally shielded from non-enzymatic browning or discolouration. Its efficacy is influenced by the fruit juice's acidity, pH, temperature, and other constituents. Because SO_2 has a bleaching effect, it should not be used in naturally coloured juices as phalsa, jamun, pomegranate, and strawberry. Additionally, as it might react with the tin of the containers, creating hydrogen sulphide and dark chemicals, it shouldn't be used in juices that will be put in tin containers. In order to preserve the juice, potassium meta-by-sulphite must first be dissolved in a tiny amount of water before being added.

Table 2: Limits of permitted preservatives in fruit juices and beverages as per FPO

Fruit juice/beverages	Preservatives	Part per million (ppm)
Fruit pulp or juice for conversion into jams and other products	SO_2	1000-3000
Fruit juice concentrate	SO_2	1500
Squashes, crushes, fruit syrup, sharbats, cordials, fruit juice and barley water	SO_2 or Benzoic acid	350 600
Sweetened ready to serve Beverages	SO_2 or Benzoic acid	70 120

- **Preservation by sugar:** Normally, fruit juices with a 66% sugar content do not ferment. Microorganisms cannot flourish in fruit syrups or sharbat that have high total solids (65% and higher). By osmosis, sugar serves as a preservative and prevents microorganism development. Jams and jellies need to be protected from mould and yeast development, which can happen on the surface, by using airtight packaging or by wrapping the item in melted paraffin wax.
- **Preservation by freezing:** Juice that is properly preserved keeps its freshness, colour, flavour, and scent for a longer time. Juice that has been deaerated is put into freezer-safe containers that have been hermetically sealed. Fruit juice cannot be sterilised by freezing, just lowers the temperature to a level where microorganisms are unable to proliferate and only very slowly undergo chemical changes.
- **Preservation by drying:** Microorganisms need water to develop and reproduce, and as soon as the water is gone, they can no longer reproduce. Fruit juices can be stored in powder form using a variety of techniques, including spray drying, drum drying, freeze-drying, foam-mat drying, etc. Fruit juice powders need to be packed with

extra care since they are quite hygroscopic. To increase the product's shelf life, it should be packed as a freely flowing powder in hermetically sealed containers with desiccant. The drying method on a foam mat may also be used to create lemon juice powder.

- **Preservation by Carbonation:** To prepare water, juice, or other liquids for serving, carbonation is the act of adding carbon dioxide under pressure. Fine bubbles that have a particular flavour are produced when the gas is released. 1 to 8 g/l of carbon dioxide is generally present in bottled carbonated beverages.
- **Preservation by Filtration:** This process involves passing clear fruit juices-particularly apple and grape juices-through specialised filters that may capture yeast and germs. This technique is utilised for soft drinks, fruit juices, and wines and involves the use of several kinds of germ-proof filters.

Filling and Processing

The bottles are thoroughly washed with hot water and filled with 1.5-2.5 cm headspace. Crown corks are used to close bottles sealed for heat processing, whilst PP (Pilfer Proof) caps can be used to seal bottles covered with chemical preservatives.

Preparation and preservation of fruit beverages

1) **Squash:** Squash is a fruit-flavored non-alcoholic concentrated syrup produced from fruit juice, water, sugar, or a sugar alternative. This is a commercially available fruit beverage that has at least 25% fruit juice or pulp and 40 to 50% total soluble solids. It also has roughly 1% acid, 350 ppm sulphur dioxide, and 600 ppm sodium benzoate. It is diluted before being served. Commercially, mango, orange, and pineapple are used to make squash. It can also be made with bael, papaya, lemon, and other fruits using potassium metabisulphite (KMS) as a preservative, or with passion fruit, peach, jamun, plum, grapefruit, raspberry, strawberry, and other fruits using sodium benzoate as a preservative. Because potassium metabisulphite can bleach anthocyanin colours, it is rarely used in dark-colored fruits. Sodium benzoate is used in these drinks.

Flow chart for processing of Squash

Fruits → Washing → Trimming → Cutting or grating → Juice extraction → Straining → Juice measuring → Preparation of syrup (Sugar + water + acid, heating just to dissolve) → Straining → Mixing with juice → Addition of preservative (0.6g KMS or 1 g sodium benzoate/liter squash) → Bottling → Capping → Storage.

2) **Ready-to-serve (RTS):** This is a sort of fruit beverage that includes at least 10% fruit juice and 10% total soluble solids, as well as roughly 0.3% acid. Because it is not diluted before serving, it is referred to as ready-to-serve (RTS). Commercial RTS drinks (containing 13% TSS and 0.3% acid) may be made with SO2 at 70 ppm or benzoic acid at 120 ppm. Because it is not diluted before serving, it is referred to as a ready-to-drink beverage. Commercially accessible goods include mango, guava, and pineapple beverages. Sweetened fruit juice is a ready-to-drink beverage with a minimum juice content of 85% and a TSS of 10%, whereas unsweetened juice is a natural juice with 100% fruit juice and natural total soluble solids. Sweetened apple and pear juices are commercially available.

Flow chart for processing of RTS beverages

Fruits (Pulp/juice) → Mixing with strained syrup solution (Sugar + water + acid, heating just to dissolve) → Homogenization → Bottling → Crown corking → Pasteurization (at about 90°C) for 25 minutes → Cooling → Storage.

3) **Cordial:** It is a clear, sparkling, sweetened fruit juice free of pulp and other insoluble elements. It comprises at least 25% juice and 30% TSS. It also includes roughly 1.5% acid and 350 ppm sulphur dioxide. This is excellent for combining with wines. Cordial can be made using lime or lemon.

Flow chart for processing of Cordial

Fruits → Washing → Cutting into halves → Juice extraction → Straining → Addition of preservative (KMS 1g/L juice) → Storing of glass container for 10-15 days for clarification (suspended materials settle down) → Symphonizing off the supernatant clear juice → Straining and measuring → Preparation of syrup → Straining and mixing of juice with syrup → Addition of preservative (0.6g KMS /liter product) → Bottling → Capping → Storage in cool and dry places.

4) **Nectar:** Nectar is prepared from fresh or concentrated fruit pulps, sugar, and an acidic ingredient. It has a high concentration of fruit pulps, fibres, and pectin, which may be consumed together with the juice. Because nectar juice has some solid substance from fruit peels and pulps, it preserves more of the fruit's original flavour, colour, and

nutrients. This fruit beverage comprises at least 20% fruit juice/pulp, 15% total soluble solids, and around 0.3% acid. Prior to serving, it is not diluted. Because it is exclusively preserved by heat processing, no preservatives are permitted in fruit nectars. This category includes commercial items such as mango nectar and guava nectar.

Flow chart for processing of Nectar

Fruits → Preliminary operations (sorting, peeling, washing and dicing) → Blending → Pasteurization (70 °C, 5mins) → Puree → Storage (-18 °C, 48hours) → Addition of sugar and adjustment (pH adjustment) → Concentration → Nectar → Addition of foam stabilizing agent → Whipping → Stabilized nectar.

5) **Syrup:** This particular fruit beverage has the following ingredients: 25% fruit juice (or) pulp, 65% TSS, 1.3 to 1.5% acidity, 350 ppm So2 or 600 ppm KMS. Prior to serving, it is diluted. Fruits such as aonla, jamun, pomegranate, grape, lemon, orange, and occasionally ginger can be used to make syrup. Additionally, rose, sandalwood, and other extracts are used to create it.

Flow chart for processing of Syrup

Fruits → Preliminary operations (sorting, peeling, washing and dicing) → Extraction of Juice → Straining → Addition of sugar syrup and citric acid → Mix the juice to hot syrup → Filling in the sterilized bottles → Pasteurization → sealing → cooling → storage.

6) **Crush:** A fruit can be crushed or squeezed to create fruit crush, a pulpy drink that is refreshing. A chilled or room-temperature serving of this beverage is offered. There are several tastes in these beverages, including strawberry, mango, and pineapple. This particular fruit drink has a minimum of 25% fruit juice or pulp and a total soluble solid content of 55%. It is diluted before consumption, has a flavor that is somewhat akin to squash, and has just 1% acid.

Flow chart for processing of Crush

Fruits → Preliminary operations (sorting, peeling, washing and dicing) → Crushing → Cooking of fruit pulp (Crush) → Addition of other Raw Materials in mixing tank (Pulp, Sugar, water, thickener, preservative, acidity regulator) → Homogenization → Pasteurization (95-98°C for 10-30 sec) → Hot filling in bottles → Storage in Dry place.

7) **Concentrate:** Concentration is the extraction of water from juice in order to enhance the level of concentration of soluble solids in fruit

juice. Spray drying or drum drying is also employed to pre-concentrate the juice for later usage. Concentrates for processing offer substantial benefits, such as lower storage, packing, and shipping costs, and the concentration of soluble solids assists in preservation by lowering water activity. Juice concentrate may also be used to make a variety of foods and beverages. Evaporation may be used to concentrate the juice, and reverse osmosis can be used to freeze the concentrated juice.

Flow chart for processing of Concentration

Fruits → Preliminary operations (sorting, peeling, washing and dicing) → Juice extraction → Heat treatment → Fine pulping (refining) → Clarified juice (Juice clarifier) → Juice concentration (Evaporator) → Juice Concentrate (at 65° Brix) → Pasteurization → Can packing (Concentrate) → Juice concentrate.

8) **Fruit juice powders:** Evaporation is used to remove water from the fruit juice during the powdering process. Because these powders are very hygroscopic in nature, they must be properly packaged. Drying processes such as spray drying, double-drum drying, vacuum drying, freeze-drying and foam-mat drying are used to make these powders from strained fruit liquids. The powder is stable and soluble in cold water. Full-strength fruit juice drinks can be made by reconstituting the powder. Much of the characteristic fresh flavour is lost during the drying process, which is compensated for by the addition of powdered natural fruit flavourings. It is possible to dry and powder mango, orange, lemon, guava, passion fruit, and tomato juice.

Advantages of Concentration and Evaporation:

- Less weight than fresh juice and reduced expenses in bulk packing, storage, and transportation.
- During the peak season, the whole production of fruits may be fully utilised, assisting in the reduction of prices.
- The product may be used to make a variety of drinks.

Advantages of unfermented fruit beverages

- **Nutritional benefits:** Unfermented fruit drinks are high in vitamins, minerals, and antioxidants, which promote general health and well-being. They can enhance digestion, stimulate the immune system, and support healthy skin and hair.

- **Low in calories:** Because unfermented fruit beverages are frequently low in calories, they offer a better alternative to sugary drinks or alcohol. This makes them an excellent alternative to maintaining or reducing weight.
- **Versatile:** Unfermented fruit drinks may be produced with a wide range of fruits and additives, giving them a versatile alternative for any taste or dietary requirement. They can be consumed alone or as a foundation for cocktails, smoothies, and other mixed beverages.
- **Convenient:** Unfermented fruit drinks are simple to prepare and can be made ahead of time for portable drinking. They're also frequently accessible in supermarkets, cafés, and juice bars.
- **Hydrating:** Unfermented fruit drinks are a great option to remain hydrated, especially in hot weather or after physical activity. They are a tasty and refreshing alternative to regular drinking water.

Disadvantages of unfermented fruit beverages

- **High in sugar:** Unfermented fruit drinks may include a lot of sugar, especially if they also have honey or agave nectar as additional sweeteners. Sugar consumption that is excessively high can lead to weight gain as well as other health problems including diabetes and tooth damage.
- **Low in fiber:** Unfermented fruit drinks, unlike entire fruits, are frequently deficient in fibre. Fibre aids digestion and can help to control blood sugar levels and decrease cholesterol. Unfermented fruit drinks may not be as full or nutritious as eating entire fruits since they lack fibre.
- **May contain additives:** Preservatives, artificial flavours, or colours may be present in certain unfermented fruit drinks. These substances can be hazardous to one's health, particularly if one has sensitivities or allergies.
- **May be expensive:** Unfermented fruit drinks might cost more than other types of beverages depending on the kind and grade of the fruit used. This may restrict their use or limit their accessibility to some persons.
- **Limited shelf life:** Fruit drinks that have not undergone fermentation are best enjoyed when they are still fresh since they have a shorter shelf life than other types of beverages. For individuals who want to stock up on beverages, they might not be as practical or handy.

Conclusion

Drinks made from unfermented fruit are a flexible and nutritious solution for individuals looking for something tasty and reviving. These non-alcoholic drinks may be tailored to meet specific dietary requirements and tastes. They can be created from a range of fresh fruits or fruit concentrates. Unfermented fruit drinks are a delightful and nutrient-dense method to remain hydrated and improve overall health owing to their high nutritional content and potential health benefits. Unfermented fruit beverages are a delightful and handy method to improve fruit consumption and enjoy the natural tastes of fresh produce, whether they are consumed on their own or as the foundation for other drinks.

References

1. Alzamora S.E, Tapia M.S, and López-Malo A. Minimally Processed Fruits and Vegetables: Fundamental Aspect and Applications. Aspen Pub. Co., Inc., Maryland, US., 2000.
2. Ashurst P.R. 1995. Production and Packaging of Non-Carbonated Fruit Juices and Fruit Beverages. Springer New York, NY.
3. Charanjit Kaur and Kapoor, H. C. (2005). Minimal Processing. Block 4 Processing and Preservation. BPVI-002, Principles of Post-Harvest Management of Fruits and Vegetables. Published by IGNOU, New Delhi.
4. Philip R. Ashurst, (2016). Chemistry and Technology of Soft Drinks and Fruit Juices, 3rd Edition. Wiley-Blackwell.
5. Gabe Carr, (2018). Fruit Production and Processing Technology. ED-Tech Press.
6. Rajauria G, & Tiwari B.K. 2018. Fruit juices: an overview. *Fruit Juices*, 3-13.
7. Rupasinghe H.V, & Yu L.J. Emerging preservation methods for fruit juices and beverages. *Food additive*, 2012: *22*, 65-82.
8. Ruxton C.H, & Myers M. Fruit juices: Are they helpful or harmful? An evidence review. *Nutrients*, 2021; *13*(6), 1815.
9. Sharma S.K. and Nautiyal M.C. Postharvest technology of Horticultural crops. New India Publishing Agency, New Delhi, 2009.
10. Siddappa G. and Tandon D.K. Preservation of Fruits and vegetables. ICAR, New Delhi, 1998.
11. Srivastva R.P. and Kumar S. Fruit and vegetable preservation: Principles and Preservation, CBS Publishers & Distibuters Pvt. Ltd., 2019.

12. Thomas Ohlsso and Bengtsson Nils, Minimal Processing Technologies in Food Industry CHIPS, 2004.

13. Verma L.R. and Joshi V.K. Post-harvest technology of fruits and vegetables. Volume 1 and 2. Indus Publishing House, New Delhi, 2000.

14. https://egyankosh.ac.in/bitstream/123456789/12106/1/Experiment-1.pdf

15. https://agritech.tnau.ac.in/postharvest/pht_fruits_intro.html

16. https://www.horticultureguruji.in/unfermented-beverages/

17. http://ecoursesonline.iasri.res.in/mod/page/view.php?id=17073

Chapter - 5

Value Addition of Fruit and Vegetables

Authors

Sanjay Kumar

Ph.D. Scholar, Department of Horticulture, CCS Haryana Agricultural University, Hisar, Haryana, India

Ankit Gavri

Ph.D. Scholar, Department of Horticulture, CCS Haryana Agricultural University, Hisar, Haryana, India

Pooja Pahal

Ph.D. Scholar, Department of Vegetable Science, CCS Haryana Agricultural University, Hisar, Haryana, India

Renu

Ph.D. Scholar, Department of Vegetable Science, CCS Haryana Agricultural University, Hisar, Haryana, India

Chapter - 5

Value Addition of Fruit and Vegetables

Sanjay Kumar, Ankit Gavri, Pooja Pahal and Renu

Abstract

Fruits and vegetables are significant dietary supplements for humans, as they provide necessary vitamins and minerals for good health and protection against various diseases and disorders. However, highly perishable nature of fruit and vegetable not allow them to store for longer period as raw produce; in addition to this market surplus during the peak harvesting season, improper and careless handling, poor storage facilities, and unregulated marketing lead to heavy losses of horticultural produce. To make available horticultural produce throughout the year, value addition of raw produce is the only way. Value addition refers to the process of enhancing the value of a product by transforming it into a form that is more attractive to consumers and can fetch a higher price in the market. Value addition can include techniques such as sorting, grading, packaging, preserving, and transforming the raw produce into processed products, such as juice, jam, pickles, or dried fruits. Value addition also involves developing and maintaining the market by identifying consumer preferences, increasing shelf life, improving packaging, and promoting the health benefits of horticultural crops. Value addition can significantly increase the economic benefits of horticultural crops for farmers, traders, and other stakeholders in the supply chain. Additionally, it can help to reduce post-harvest losses, increase food security, and improve the overall quality of life for communities involved in horticultural production.

Introduction

India had rapid expansion in the food processing industry after post-independence, particularly in the early 1980s. The increase in agriculture production after the green revolution resulted in need of its postharvest management. The agripreneurs recognised the importance of the post-harvest or processing industry, which led to a shift from producing agricultural goods to processing (Kachru, 2006). However, in case of horticulture sector, consumer prefers fresh produce over processed products particularly for fruit and vegetable. In such situation, the processing industry has been unable to fulfil the requirements sufficiently. Since the Green Revolution, which primarily focused on increasing food grain production to address the nation's food security issue. The shift in cropping patterns over the last ten years has been more in favour of commercial crops and the horticultural industry (Mittal, 2007). Presently, horticulture crops are grown in an area of 25.4 million with 311.71 annual production (NHB 2018). India's nutritional security is greatly influenced by horticultural crops, which include fruits, vegetables, flowers, medicinal and aromatics plants, spices, and plantation crops. Even though they only cover 8.5% of arable land, these crops account for 29.5% of agriculture's gross domestic production (Economic Survey 2007-2008). This need technologically driven development. Horticultural crops provide an ample employment opportunities to rural people, as these crops are highly labour intensive.

Fruits and vegetables are significant dietary supplements for humans, as they provide necessary vitamins and minerals for good health and protection against various diseases and disorders. Human dieticians suggest that an adult who works moderately required 300 g of vegetables and 85 g of fruits in addition to grains, fish, meat, and milk per day. Fortunately, India's vast climatic condition and its wide range of soil provide an ample opportunity to grow a variety of horticultural crops and are emerged as a second largest producer of fruits and vegetables in the world. The nation is now emerged as the second largest producer and exporter of tea, coffee, and cashews in the world and the top producer of spices, coconut, and tea. According to Nanda *et al.* (2012), although food grains have been taken care of in the post-Green Revolution era, horticulture commodities, primarily fruits and vegetables, have experienced postharvest losses due to the necessity for basic processing, preservation, and transport technology. Although horticulture production has performed well, India's contribution to value addition has fallen short of expectations. India accounts only 2% of the global horticultural trade. Although there may be a market surplus during the peak harvesting season, improper and careless handling, storage, and marketing lead to physical damage from tissue breakdown. Due to poor transportation services and a lack of packaging materials availability, the surplus produce cannot be supplied to markets in urban areas. Furthermore, due to a lack of suitable local cold storage facilities, the surplus produce is frequently unable to be kept for sale during the off-season. In such situations, the growers do not receive a fair price for their crop due to the market glut, and spoilage of produce. About 10–15 % fresh farm produce shrivel and decay, reduce their market value and consumer acceptability. Value addition of fruits and vegetables can assist to address these issues. Also, the processing industry can utilize undesirable fruits and vegetables that are small, misshapen fruit, overripe, and diseased produce. Demand for industrially produced goods is rising in the domestic market as a result of increased urbanisation, rising middle-class purchasing power, changes in eating patterns, and a decrease in the practise of preparing preserved goods at home. Reducing these losses can boost their supply and enhance the general nutrition of the average Indian without adding more land to the agricultural sector. The losses in farm products in India have been estimated to be of a very high order mainly due to poor postharvest management. According to studies, by adopting the right agro-processing technology might reduce losses to less than 50% of their current level. It is necessary to call new initiatives in order to reduce the remaining losses. In order to prevent losses, long-term attention should be given to appropriate farm produce storage buildings, cold stores, and processing equipment. The more products we preserve, the more we will be able to consume in the future as value addition takes care of extra produce and prevents wastage from decomposing. By reducing wastage amid a market glut, value addition or processing helps farmers get a higher return. It preserves the products in a proper way that is impossible under normal or cold storage settings. Therefore, value addition or processing is a good alternative to storage. It enables the availability of fruits and vegetables during the off-season and in areas where they are not cultivated. Additionally, produce that have been preserved are more appetising.

In last few years, horticulture shifted more towards processing industry by producing many products like various fermented and non-fermented beverages, pulps, dehydrated and canned fruit and vegetable products, pickles, pastes, and curried vegetables. Furthermore, Indian government introduces various schemes, subsidies and the concept of agri-export zones and mega food parks to boost the food-processing industry and also its subsector like fruit and vegetable processing industry in India. India exported 1,586,255.05 million tonnes of processed fruits and vegetables valuing 1,285,866.02 lakh rupees (APEDA 2021-22).

The introduction of large-scale export-oriented facilities for the production of value added product such as sorting, grading, packaging, preserving, and transforming the raw produce into processed products, such as juice, jam, pickles, or dried products of raw fruits and vegetables still requires the import of advanced technology.

Advantages of value-added products

- **Longer shelf life:** Processing fruits and vegetables into value-added products can extend their shelf life, making them more stable and reducing the chances of spoilage. This increases their marketability and profitability.

- **Greater convenience:** Value-added products are often more convenient for consumers to use, as they are pre-prepared and can be consumed or cooked with minimal preparation.
- **Increased variety:** Processing fruits and vegetables into value-added products allows for greater variety and diversity of products, as different flavors, textures, and ingredients can be added.
- **Higher profit margins:** Value-added products typically command a higher price than raw fruits and vegetables, which can lead to higher profit margins for producers and processors.
- **More opportunities for small-scale producers**: Value-added products can provide small-scale farmers with an opportunity to add value to their produce and access new markets, which can help to increase their income and profitability.
- **Reduced waste:** Value-added products can help to reduce food waste by allowing farmers and processors to use produce that is not suitable for the fresh market or has blemishes that would otherwise go to waste.

Different value added or processed products of fruit and vegetable

Value addition refers to the process of enhancing the value of a product by transforming it into a form that is more attractive to consumers and can fetch a higher price in the market. Some of the value added or processed product of horticultural produce includes: Beverages either fermented or non-fermented, in semisolid form as pulp, jam, jelly, marmalade, chutney etc.; and in solid form such as canning, drying, candy, pickle and preserve *etc*.

1) Beverages

The term beverages refer to all fermented / unfermented, sweetened/ unsweetened drink. Fruit juices, ready to serve (RTS) drinks, cordial, nectar, syrup, wine, cider are some common examples of beverages. Presently synthetic beverages are manufactured in huge quantities and becoming common in Indian market:

A) **Unfermented beverages:** These beverages are mainly prepared from fruit juice that do not undergo alcoholic fermentation process.

i) **Fruit juice:** Pure fruit juice is a beverage made from the liquid extracted from fresh fruits. The juice can be extracted using a juicer or a blender, and it may be consumed either as is or with added water or sweeteners. Pure fruit juice contains natural sugars, vitamins, minerals, and antioxidants that are beneficial for health. E.g.: citrus juice, apple juice, *etc.*

ii) **Fruit juice beverage:** It is highly refreshing, thirst quenching, appetizing, easily digestible and nutritionally far superior to synthetic drink. It is natural extracted coarsely strained fruit juice, containing moderate quantity of fruit pulp, that is significantly altered in composition by adding water and a small amount of sugar during processing. For longer shelf life, acid and chemical preservatives may be added according to requirements. It may be further altered in composition and diluted before consumption, as pineapple and papaya.

iii) **Squash:** It is a kind of fruit beverage that containing moderate quantity of fruit juice or pulp *i.e.*, 25% and 40-50% total soluble solids (TSS), 1.0% acid, and suitable preservative. This beverage is diluted before serving (Jood and Ketarpaul, 2002). E.g.: orange squash, lemon squash, pineapple squash, *etc.*

iv) **Cordial**: A sparkling clear fruit juice extracted from fruit, from which all the pulp and suspended materials are removed completely. It may be sweetened by adding sugar. As per F.P.O. specification it must contain 25% fruit juice, 30% TSS, 1.5% acid, and preservative. E.g.: lime, orange, *etc.*

v) **Crush:** A fruit beverage containing at least 25% fruit juice or pulp and 55% TSS. It also contains 1.0% acid, and is diluted before use, such as pineapple crush.

vi) **Fruit juice concentrate:** It is concentrated fruit juice mainly prepared by removal of water by various methods as evaporation, reverse osmosis, heat or freezing *etc.* Carbonated beverages and other products are made from the concentrate, which contains at least 32% TSS.

vii) **Ready-to-serve (RTS):** This fruit beverage contains at least 10 % fruit juice, 10 % TSS and about 0.3% acid. It never diluted before consumption. It is prepared from ber, jamun, guava, *etc.*

viii) **Nectar:** It refer to fruit beverage that is coarsely strained and containing at least 20% fruit juice/ pulp, 15% TSS, and about 0.3% acid. It is not diluted before use. Jamun, bale, guava are few examples.

ix) **Syrup:** As per F.P.O. specification, it is a kind of beverage containing 25% fruit juice or pulp, 65% TSS and 1.3-1.5% acid. It is diluted before serving (Sharma *et al.,* 1988). E.g.: pomegranate, jamun, pineapple, strawberry, etc.

x) **Synthetic syrup or sarbat:** It is prepared by using concentrated sugar syrup of 70-75 % strength with the utilization of fruit, herbs as a source for flavouring and colour.

xi) **Barley water:** Fruit beverage that contains 0.25% barley starch along with 25% fruit juice, 30% TSS. Barley water is exclusively prepared from citrus fruits such as lime, lemon, grapefruit, and orange. Lime and lemon are most preferred fruit for preparation of barley water.

xii) **Carbonated beverage:** These beverages are prepared by using carbon dioxide gas as preservative; it is then called a carbonated beverage. Orange juice preserved in this method is known as orangeade, and, similarly, lemon juice known as lemonade.

B) **Fermented beverages:** The beverages prepared from alcoholic fermentation of fruit juices by yeasts. Wine, champagne, port, sherry, tokay, muscat, perry, orange wine, berry wine, neera, and cider are some of the popular example of fermented beverages.

i) **Wine:** It is sparkling clear liquid prepared by fermentation of fruit having 10-12% juice content with yeast (*Saccharomyces ellipsoideus*, S. *malei*, S. *cerevisiae*) in anaerobic conditions and at 25-27 °C, after its sterilization ethyl alcohol is produced which is then filtered and stored in an airtight container to check further infection and thereby fermentation with vinegar bacteria. It can be classified based on the alcohol concentration such as light wine that is having 7–9 % alcohol content; medium wine (9-16 % alcohol content and strong wine (16-20 % alcohol content). Further 4-6 % alcohol content present in fruit brandy. Wine mainly prepared from fruit as grape wine. It is also classified based on the concentration of sugar as dry wines which contain practically little or no sugar; and sweet wines that contain some sugar and have a sweeter taste.

ii) **Champagne:** Champagne is sparkling clear wine especially popular in France by using certain varieties of grapes.

iii) **Port:** It is fortified sweet red wine indigenous to Portugal, but it is now becoming popular in other countries as well.

iv) **Tokay:** This is a famous fortified wine from Hungary.

v) **Sherry:** Sherry is a Spanish wine that is mainly prepared from white grapes.

vi) **Perry:** It especially prepared from pear fruit by using the wastage, culled fruits, and trimmings left over from canaries.

vii) **Feni:** Feni is a type of alcoholic drink that is popular in the Indian state of Goa. It is made from cashew apples, which are crushed and then distilled to produce strong liquor. It is registered as geographic indication tag of Goa for its distinctive taste and aroma (Serkar and Mariappan, 2007).

viii) **Neera:** It is a non-alcoholic and nutritious drink prepared from coconut. The vascular sap collected from immature unopened coconut inflorescence is popularly known as 'Neera' in fresh form.

ix) **Vinegar:** The earliest known fermentation product is vinegar. It is obtained by fermenting various ingredients that contain sugar and starch with alcohol and acetic acid. It has a distinctive smell and scent because to the presence of about 5% acetic acid in water, as well as various amounts of fixed fruit acids, colouring agents, salts, and other fermentation by-products. In the market, vinegar is identified by the substance from which it was created. Vinegar made from malt is known as malt vinegar, apple cider vinegar is known as apple juice vinegar, and grape vinegar is known as grape vinegar.

2) Semisolid Form

a) **Jam:** It is a product prepared from fruit pulp after cooking with sufficient quantity of sugar to a reasonably thick consistency. Generally, it is not transparent because it holds fruit pulp in its composition. As per F.P.O specification jam must contain 68% TSS, 45% portion of fruit pulp and 0.5-0.6% acid content. A good jam neither syrupy nor stiff rather, it must have bright colour along with true fruit flavour and well distributed fruit portion. It should be free from crystallization of sugar. Fruit suitable for jam making are mango, pineapple, apple *etc*.

b) **Jelly:** It is semisolid transparent product prepared from clear extract of pectin containing fruit after cooking with sufficient quantity of sugar. Jelly contains 0.50-0.75% acid, 65% TSS and 45% of fruit juice. A perfect jelly should be non-sticky, non-syrupy and non-gummy. It must be transparent; firm enough to retain shape edge even cut with a knife and have strong flavour of the original fruit. Guava is most suitable fruit for jelly preparation.

c) **Marmalade:** This product is almost similar to jelly, but it prepared from mainly citrus fruit (sweet orange, lemon, grapefruit, *etc*.) after cooking pulp or juices with sufficient sugar and using shreds of peel as a suspended material.

d) **Chutney:** The term chutney refer to a product having reasonably thick consistency prepared from matured green fruits or matured but tender vegetables by peeling, boiling, crushing, mixing and cooking with acid, sugar, salt, coarsely powdered spices, and herbs. In chutney the higher content of sugar, acid, spices, or vinegar jointly act as preservatives. E.g.: mango, tomato.

3) **Solid Form**

a) **Canning:** Canning refers to filling of food stuff in air tight sealed container followed by sterilizing the container by heat for longer storage. Generally, in case of fruit, syrup having sugar concentration about 30-50% is used for dipping the whole fruits or pieces of fruits; similarly for vegetable brine solution having salt concentration 3-5 % is used for canning. Also blanching should be practiced before canning in case vegetables. Almost all the fruit and vegetables may be stored in can.

b) **Drying:** One of the earliest techniques used to preserve food is drying. Food spoilage organism require a minimum moisture content in produce to feed, develop, and proliferate in order to cause food to decompose. When a produce is dried, the moisture content is lowered to a point where the growth and development of food spoilage microbes inhibited completely. It is mainly achieved either by sun drying or under controlled temperature and humidity conditions inside an oven. Some important dried or dehydrated products are Fig, Resins, dehydrated ber, apricot *etc*.

c) **Leather:** Leather produced by drying of strained pulp of fruits or vegetable with addition of small amount of sugar then spreading on an alumina plate until dried. It may be dried in a thin layer in a solar or cabinet drier. This process of spreading fruit pulp on previously dried pulp repeated until the thickness reached 2-3cm. Then after it is fumigated in sulphur chamber and placed by wrapping in butter paper. E.g.: mango leather, tomato leather, papaya leather, custard leather, jackfruit leather, palm leather, jamun leather, *etc*.

d) **Flake:** It refer to a single layered fully dried pulp of fruit and vegetable directly collected from an aluminium plate as dry thin pieces. E.g.: tomato flake, potato flake, corn flake, *etc*.

e) **Preserves:** A product prepared from whole or pieces of properly matured fruit pieces or mature and tender vegetable pieces after impregnating with heavy sugar syrup till it become tender and transparent. A good quality preserve must contain at least 60 % of soluble solid. E.g.: Aonla preserves, Bael preserves, Karonda preserves *etc*.

f) **Candy:** It is a product prepared after draining and drying of sugar syrup from preserve. In this prepared preserve is placed over wire mesh for draining out of sugar syrup and this result in preparation of candy. Further thin transparent sugar coating is covered over candied fruit imparts glossy appearance to produce is called glazed fruits. E.g.: ber, aonla, *etc*.

g) **Pickles:** It is one of the earliest methods of fruit and vegetable processing. As pickle are good source of appetizer and add palatability in food. Several kind of pickle marketed in India. Among various kind of pickle mango pickle rank first, followed by cauliflower and others. Pickles are commonly manufactured in home as well as in commercial. Preservation of fruit or vegetable whole or its pieces using common salt, oil, spices, vinegar or mixing of all these is termed as pickling. It is prepared by fermentation of fruit or vegetable whole or its pieces with salt having concentration 8-10 % and 3 % acid under sun exposing at 29.4-30 ^{0}C temperature for 6-8 days with the help of lactic acid bacteria. Further it thoroughly mixed with coarsely spices powdered with or without vinegar and further, sealed by covering fully with moisture-free edible oil to produce pickle. Pickles should be kept another 3 weeks in the sun. E.g.: lime, mango, cauliflower, ber, *etc*.

Some important value added products of fruit and vegetable crops

Sauces and Ketchups: It is one of the most popular value added product prepared from tomato. But now a day's various fruit and vegetable are used for preparation of sauces and ketchups. For preparation of Sauces and Ketchups fruit or vegetable pulp or juice is sieved after cooking to remove the skin, seeds, and stalks of the fruits, vegetables, and spices and to give a smooth consistency to the final product. The method of preparation and ingredients are same as chutney. There is no essential difference between sauce and ketchup. However, sauces are generally thinner and contain more total solids *i.e.*, 30% than ketchups 28%. Tomato, apple, papaya, walnut, soybean, mushroom, *etc.*, are used for making sauces.

Potato Chips: It is one of the most famous value added potato product. Generally fresh potatoes which contain high dry matter content are preferred for potato chips. Pandey *et al.* (2005) reported that cv. Kufri Chipsona-1 exhibited high dry matter content and low reducing sugars is suitable for chips making. Kufri Chipsonna-3 and Kufri Chipsona-4 are recommended varieties for making potato chips.

French Fries: It is the famous snack typically prepared from deep fried potato that has been cut into various shapes, especially thin strips. Kufri Frysona and Kufri Chipsona-3 are suitable varieties for preparation of French fries.

Sauerkraut: It is most popular processed product of cabbage in countries like Europe and the USA. The word Sauerkraut represents acid cabbage. Sauerkraut is a traditional dish made by fermenting shredded cabbage with lactic acid bacteria. It is considered to be a healthy food because it is low in calories and high in fiber, vitamins, and minerals.

Petha: It is made by cooking pieces of ash gourd in sugar syrup until they become translucent and candy-like. Petha can be flavoured with various ingredients such as saffron, rose water, or cardamom, and is often served as a dessert or snack. It is a traditional specialty of the city of Agra.

Sago: It is partially dried starch cake of cassava. It is mainly used as food for new born babies and preparation of pudding and sago wafer (Nanda, 2008).

Coconut Toddy: Coconut toddy is a popular alcoholic beverage made from the fermented sap of coconut trees. The sap is collected by tapping the flower bud of the coconut tree, and then the sap is collected in a container. The collected sap is then fermented naturally for several hours or days, depending on the desired level of alcohol content. The resulting drink is sweet and mildly alcoholic, with a tangy flavour that is unique to coconut toddy.

Chali or Kottapak: It is processed product prepared from dehusking of sun dried arecanut. Kalipak Nuts of 6-7 month old maturity are dehusked, cut into pieces, boiled with water or a diluted extract from previous boiling, coated with kali and dried.

List of some value-added products of fruits and vegetables in India

Crops	Value added products
Apple	Cider, juice, jam, pulp, osmotically dried rings, canned apple, carbonated juice, fibre from pomace
Apricot	Pulp, squash, RTS, jam, dried apricot, osmotically dried apricot, apricot oil based cream
Plum	Pulp, squash, RTS, jam, chutney, wine/brandy, seed oil
Peach	Canned peach, pulp, jam/chutney, plum sauce, kernel oil
Pear	Canned pear, pulp, jam, sand pear candy, vermouth
Mango	Pulp, RTS, squash, amchur powder, pickle, leather
Grapes	Raisins, wine, juice
Litchi	Juice, squash, RTS/nectar, carbonated drink
Cauliflower	Pickle, slices in brine, frozen cauliflower head
Cabbage	Sauerkraut
Carrot	Pickle, preserve, candy, kanji
Chillies	Green chilli puree, powder, green chilli paste, oleoresin
Tomato	Juice, sauce, ketchup, paste, puree
Pea	Dehydrated pea, frozen pea
Potato	Chips, French fries, puffs, flour, flake, wine
Black pepper	Oleoresin, green pepper in brine, dehydrated green peppers, canned green pepper, pepper oil, freeze-dried green pepper, white pepper powder

Constraints and Opportunities in Value Addition of Horticultural Crops in India

The value addition of horticultural crops is important as it can lead to increased income for farmers, promote food security, and create employment opportunities. However, there are several constraints that can limit the effectiveness of value addition in horticulture. Some of these constraints include:

Lack of infrastructure: The value addition process requires adequate infrastructure such as processing plants, storage facilities, and transportation networks. In many developing countries, such infrastructure is lacking, making it difficult to effectively carry out value addition activities.

Limited technical knowledge: Value addition requires technical skills and knowledge in areas such as processing, packaging, and marketing. However, many farmers lack such knowledge, which limits their ability to add value to their produce.

Quality issues: Value addition requires high-quality produce, but the quality of horticultural crops can be affected by factors such as pests and diseases, poor handling, and inadequate post-harvest management practices. This can make it difficult to produce high-quality value-added products.

Limited access to markets: Value-added products often require access to niche markets that are willing to pay a premium for high-quality, processed products. However, many small-scale farmers lack access to such markets, which limits their ability to sell their products at a higher price.

Lack of finance: Value addition requires significant investment in infrastructure, equipment, and marketing. However, many small-scale farmers lack the necessary financial resources to make such investments.

Policy and regulatory constraints: Policies and regulations can also limit the effectiveness of value addition in horticulture. For example, high taxes and tariffs can make it difficult to import the necessary equipment and raw materials, while restrictive regulations can limit the development of new value-added products.

Opportunities

1) Contract farming system must be introduced for assured supply of raw material to processing industry and for assuring famer income.
2) Advanced postharvest technologies should be introduced to improve the handling, packaging, transportation and storage of horticultural produce.
3) Low-cost storage structures should be introduced for the horticultural produce at farmer field and despatch areas.
4) Development of complete cool chain for maintaining the desirable quality of the perishable produce from farm to final consumption.
5) Non-thermal/non-chemical processing should be adopted to minimize the nutritional alteration in food products.
6) Recycling of horticultural wastes specially fruit, vegetable and plantation crops wastes into newer products.

Literature cited

1. APEDA. 2021-22. Promoting fruit and vegetable consumption around the world. http://www.apeda.gov.in/apedawebsite. Accessed 11 May 2023.

2. Jood S, Ketarpaul, N. 2002. Food preservation. Agrotech Publishing Academy, Udaipur.

3. Kachru R.P. 2006. Agro-processing industries in India growth. Status and prospects. Indian Council of Agricultural Research, New Delhi.

4. Mittal S. Can horticulture be a success story for India? Working paper no. 197. Indian Council for Research on International Economic Relations. ICRIER (Indian Council for Research on International Economic Relations), New Delhi, 2007.

5. Nanda S.K. Postharvest technologies and utilization. In: Palaniswami MS, Peter KV (eds.) Tuber and root crops. New India Publishing Agency, New Delhi, 2008.

6. Nanda S.K, Vishwakarma R.K, Bathla H.V.L, Rai A. and Chandra P. 2012. Harvest and postharvest losses of major crops and livestock produce in India. AICRP on PHT, CIPHET (ICAR), Ludhiana.

7. National Horticulture Board. 2018. National Horticulture Board Data Base. http://nhb.gov.in/

8. Pandey, S.K., Paul Khurana, S.M., Singh, S.V., Kumar, D. and Kumar, P. 2005. Evaluation of Indian and exotic varieties for sustaining processing industries in North Western parts of India. *Indian Journal of Horticulture*, 62(2): 155-159.

9. Serkar S. and Mariappan S. 2007. Usage of traditional fermented products by Indian rural folks and IPR. *Indian Journal of Traditional Knowledge*, 6(1): 111-120.

10. Sharma, K.D., Sethi, V. and Maini, S.B. 1988. Studies on incorporation of concentrate in covering syrup for processing of apple rings in flexible pouches. *Journal of Food Science & Technology*, 35: 371-374.

Chapter - 6

Post-Harvest Handling Operations; Cleaning: Cleaning of Grains, Washing of Fruits and Vegetables, types of Cleaners, Screens, types of Screens, Rotary Screens, Vibrating Screens, Machinery for Cleaning of Fruits and Vegetables (Air cleaners, Washers), Cleaning Efficiency, Care and Maintenance

Authors

Aayush Singla

M.Sc. Scholar, Department of Horticulture, Chaudhary Charan Singh Haryana Agricultural University, Hisar, Haryana, India

Ankit Gavri

Ph.D. Scholar, Department of Horticulture, Chaudhary Charan Singh Haryana Agricultural University, Hisar, Haryana, India

Rajvinder Singh

Ph.D. Scholar, Department of Genetics and Plant Breeding, Chaudhary Charan Singh Haryana Agricultural University, Hisar, Haryana, India

Tanu Chaudhary

M.Sc. Scholar, Department of Food and Nutrition and Food Technology, University of Delhi, Delhi, India

Chapter - 6

Post-Harvest Handling Operations; Cleaning: Cleaning of Grains, Washing of Fruits and Vegetables, types of Cleaners, Screens, types of Screens, Rotary Screens, Vibrating Screens, Machinery for Cleaning of Fruits and Vegetables (Air cleaners, Washers), Cleaning Efficiency, Care and Maintenance

Aayush Singla, Ankit Gavri, Rajvinder Singh and Tanu Chaudhary

Abstract

Every agriculture and horticulture produce requires a precise post-harvest treatment, however, there are definite common principles that has to be applied on them. After harvesting agriculture and horticulture produce undergo a number of processing stages between harvest and final consumption. This chain of processes is often stated as the total post-harvest system. The post-harvest system can be divided into three different areas. The first is the preparation of produce for storage. The second, which is referred to as primary processing, involves further treatment is to clean them, remove the husk, dirt or any other impurities or reduce the size. The products from primary processing are still not consumable. The third stage (secondary processing) transforms the produce into edible products. Primary processing includes various different procedures like cleaning, sort and remove the inedible, improper, misshapen, disease infested fractions from the grains, fruits or vegetables. Primary processing of cereals includes cleaning, grading, hulling, milling, pounding, grinding, tempering, parboiling, soaking, drying, sieving. Cleaning is one of the most critical post-harvest handling operations that help in maintaining the quality of agricultural produce. Proper cleaning ensures the removal of dirt, foreign matter and any potential contaminants that may have accumulated during harvesting, transportation and storage. This chapter provides an overview of cleaning operations, including cleaning of grains, washing of fruits and vegetables, types of cleaners, screens, machinery for cleaning of fruits and vegetables, cleaning efficiency, care and maintenance.

Introduction

In the intricate journey from field to table, post-harvest handling plays a vital role in preserving the quality and safety of agricultural produce. Among the various operations involved, cleaning and washing are paramount steps that significantly impact the shelf life and market value of harvested crops. This chapter delves into the essential aspects of post-harvest cleaning and washing operations, exploring their importance, techniques, and the influence they have on the overall post-harvest management process. As the name suggests, cleaning and washing are key processes employed to remove extraneous matter, such as dirt, debris and microbial contaminants, from harvested crops. These operations not only enhance the visual appeal of the produce but also eliminate potential pathogens and spoilage microorganisms that could compromise its safety and freshness. By effectively cleaning and washing harvested crops, farmers, processors, and distributors can extend the storage life, reduce post-harvest losses and maintain the nutritional quality of the produce (Gunathilake *et al.*, 2018).

The chapter begins by providing an overview of the objectives and significance of cleaning and washing operations in post-harvest management. It highlights the benefits of thorough cleaning and proper washing techniques in mitigating the risks associated with microbial contamination, pesticide residues, and foreign material. After washing, surplus water should be removed (Sudheer and Indira 2007). The chapter also explores the impact of cleaning and washing on different types of produce, emphasizing the specific requirements and challenges associated with various crops.

Furthermore, this chapter delves into the multifaceted realm of post-harvest cleaning, exploring its importance in the context of grains, fruits, and vegetables. Additionally, it discusses the various types of cleaners and screens employed, including rotary and vibrating screens, as well as machinery such as air cleaners and washers. Furthermore, the chapter investigates cleaning efficiency and offers guidance on the care and maintenance of cleaning equipment, allowing readers to optimize their cleaning processes for enhanced post-harvest handling outcomes (Madan and Ullasa 1993).

Throughout the chapter, attention is given to the various factors that influence the success of cleaning and washing operations. This includes the consideration of crop-specific characteristics, such as surface texture, fragility and susceptibility to mechanical damage. Additionally, the chapter highlights the importance of maintaining proper sanitation and hygiene practices in cleaning and washing facilities to prevent cross-contamination and ensure food safety.

Cleaning of Grains

Cleaning is the removal of debris and other impurities to make grains pure and free from undesired, non-consumable items. Grain after passing through the oscillating straw rack still contain sufficient amount of chaff and broken plants debris. Cleaning of grains from such impurities is either carried out by the oscillating sieving mechanism or by the pneumatic method. The latter exploits the aerodynamic characteristics of particle mixtures. Cleaning of grain by the force of wind is a prehistoric method and not really effective. But assistance of air is use in all cleaning mechanisms, (a) to remove light materials, (b) to assist in orienting the particles on the perforated sieves, and (c) to move the particles caught in the opening of perforated sieves during the oscillation process.

Importance of Cleaning:

Effective cleaning of agricultural produce is crucial for several reasons:

- **Quality Preservation:** Cleaning helps maintain the visual appeal and overall quality of the produce. By removing dirt, dust and extraneous materials. It enhances the appearance and marketability of the products.
- **Safety and Hygiene:** Cleaning eliminates potential contaminants, including bacteria, fungi, and pesticide residues, reducing the risk of foodborne illnesses and ensuring the safety of consumers.
- **Shelf-life Extension:** Properly cleaned produce is less prone to decay and spoilage. By removing organic matter that can promote microbial growth, cleaning helps extend the shelf life of the produce.
- **Processing Efficiency:** Clean produce is easier to handle during processing and reduces the risk of equipment clogging or damage, resulting in more efficient operations.

Methods of Cleaning: There are several methods for cleaning grains, depending on the specific type of grain and the desired level of cleaning. Here are some common methods used in grain cleaning:

- **Manual Cleaning:** Manual cleaning involves the physical removal of dirt, debris, and damaged parts of the produce. It is commonly used for delicate fruits and vegetables. Manual methods include hand-picking, brushing, washing with water or using soft cloths. Brushes with soft or hard bristles are used for scrubbing produce surfaces to remove dirt and debris. Bristle rollers, either manual or mechanical, are effective for cleaning root crops like potatoes and carrots.

- **Sieving:** This method involves passing the grains through a series of screens or sieves with different mesh sizes. The screens allow smaller impurities such as dirt, stones, and broken grains to fall through, while the larger, clean grains remain on top.
- **Gravity separation:** This method takes advantage of the difference in density between the grains and the impurities. By adjusting the airflow and angle of inclination, lighter impurities like chaff and dust can be blown away, while the heavier grains fall into a collection area.
- **Magnetic separation:** Magnetic separators are used to remove ferrous impurities from grains. The grains pass through a magnetic field that attracts and removes any magnetic materials such as iron particles.
- **Air aspiration:** This method uses an air stream to separate light impurities from the grains. The grains are fed into an aspirator where a controlled air flow blows away lighter materials like chaff, husks, and straw, leaving behind the clean grains.
- **De-stoning:** In this process, grains are passed through a series of screens or sieves with different hole sizes. As the grains move across the screens, any stones or heavy impurities are separated and removed.
- **Washing:** Washing is typically used for certain types of grains, such as rice. It involves soaking the grains in water and agitating them to remove dirt, dust, and other impurities. The grains are then rinsed and dried before further processing.
- **Water Cleaning:** Water is a widely used medium for cleaning fruits, vegetables and certain grains. Immersion or spray systems can be employed, depending on the produce type and desired level of cleaning. Water can be combined with mild detergents or sanitizers for enhanced cleaning and disinfection. Washers utilize water sprays, agitation and sometimes brushes to clean fruits and vegetables. Dippers, often used for delicate produce, involve immersing the produce in water and manually agitating it to dislodge impurities.
- **Dry Cleaning:** Dry cleaning methods involve using various tools and equipment to remove dirt and impurities without the use of water. Examples include brushing, vacuuming and air blowers. Dry cleaning is commonly employed for grains, seeds and certain fruits

that are sensitive to excessive moisture. Air cleaners use high-velocity air to remove dust, chaff, and lightweight particles from grains and seeds. They are particularly useful for cleaning cereal grains and legumes.

Types of cleaners

Cleaners are devices used to remove foreign materials from agricultural produce. They come in different types, including air cleaners, gravity separators, and screen cleaners. Air cleaners use a stream of air to blow off dust and chaff from the produce. Gravity separators separate the produce based on the differences in weight and density. Screen cleaners, on the other hand, use a series of screens to remove impurities based on their size.

- **Scalper:** A scalper can be used to rough clean seed when trash content is high. It consists of vibrating or rotating screen or a sieve. The screen perforations are large enough to allow the rough seed to pass through readily while large inert material is scalped off and removed from the grain lot. It provides much higher output than fine cleaning.
- **De-awner:** It consists of a totally enclosed rigid steel case equipped with an agitator. A shaft with staggered steel beaters revolves at high speed through the material, thereby removing, for example, hairs from barley and splitting double kernels, for example in wheat. It may be helpful in breaking dormancy in some species.

Scalper

De awner

- **Fanning mill:** It is type of screen cleaner that employ an air blast to assist in cleaning. Though these are used for cleaning but these are not much effective as regards to density and if the material have same shape. Here in addition to the aspirating fan used to draw air through the material fed on 1st screen unit, another fan is used to force the air through the product after fine grading.

- **Air screen cleaners:** It uses two air blasts and two screens. The first air system removes dust and light chaff for the seed reaches the first screen. The first screen allows the good seed to drop onto the second screen. The large foreign material rides over the first screen and is discarded. The second screen is a grading screen.
- **Vibratory screen cleaner:** It consists of more than one screens. All the screens are tightened together and suspended in such a manner that these have horizontal oscillating and slightly vertically motion. The slope of the screen is adjustable to control the rate of forward travel of the grain. Two rushes also attached to the screen, in order to avoid clogging. Coarse impurities are screened off y upper screen. While, fine impurities are screened off by lower screen. After that, grain again cleaned y upward draft of air in ascending separator.
- **Rotary air screen cleaner:** The rotary air screen cleaner consists of a cylindrical drum or screen that rotates horizontally. The drum is typically perforated or made of mesh, allowing air and small particles to pass through while retaining larger materials. The rotation of the drum helps to separate and sort the materials based on their size and weight. Grains come out at the centre side of the drum rotating at low speed and fall onto the vibratory screen which remove the dirt particles.
- **Specific gravity separators:** Seeds or grains fed on the tale in 3-5 seed thickness. The shaking deck pushes the heavy seeds uphill with deck and air steam float lighter seed. Slope tends to stratification while oscillation controls movement.
- **Indented cylinder separator:** It consists of indents which lift particles that fit into the indents. Particles which do not fit drop out of the cell and fall downwards.

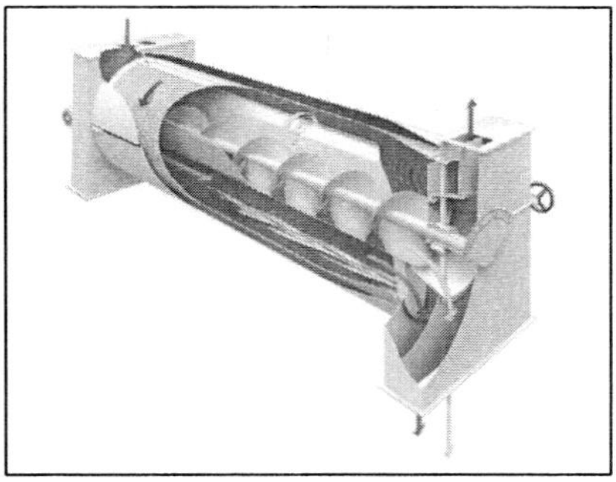

Fanning mill **Indent cylinder separator**

- **Cyclone separator:** It is used extremely in processing and other operations as a device for collecting end product and also in connection with pneumatic conveying of products and wastes from processing. These are used for removal of particle about 10 nanometre or in diameter from the air stream.
- **Pneumatic separator:** It uses aspirating column for cleaning of grains. In this machinery, grains are fed from the duct provided at the side. It uses the terminal velocity for separation of grains from another material. This is best suited for removing very small sized and immature grain from healthy grains.
- **Inclined draper:** This machine uses differences in shape and surface texture of the material for cleaning of grains. The mixture to be separated is fed over the centre of an inclined draper belt moving direction. The round and smooth grains roll or slide down the draper at faster rate than the upward motion of the belt and discharged in a hopper. The flat shape or rough shape particles are carried to the top of the inclined draper.

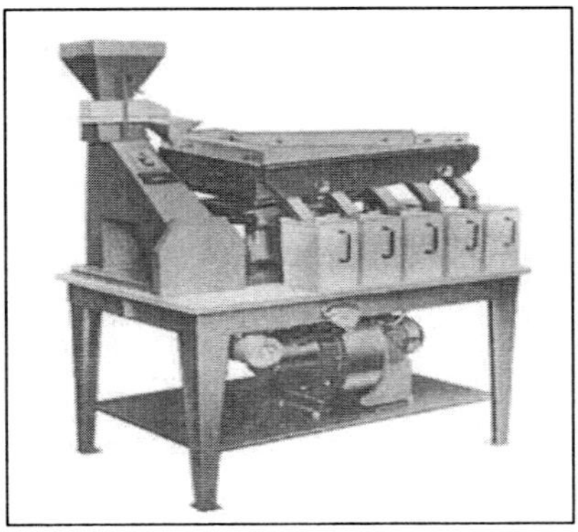

Air screen cleaner **Vibrating screen cleaner**

Cyclone separator

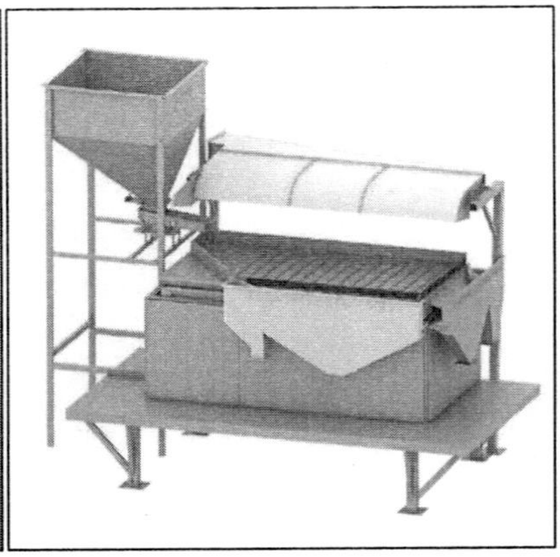

Gravity separator

- **Magnetic separator:** Magnetic separators employ magnets to attract and remove ferrous materials, such as iron particles, from grains. This method is particularly useful for removing metal contaminants that may have entered the grain during harvesting or handling.

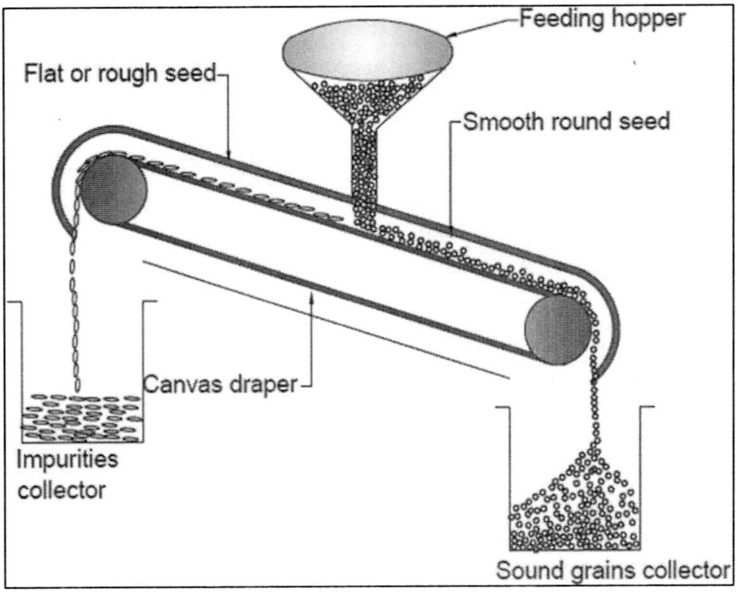

Inclined drapper

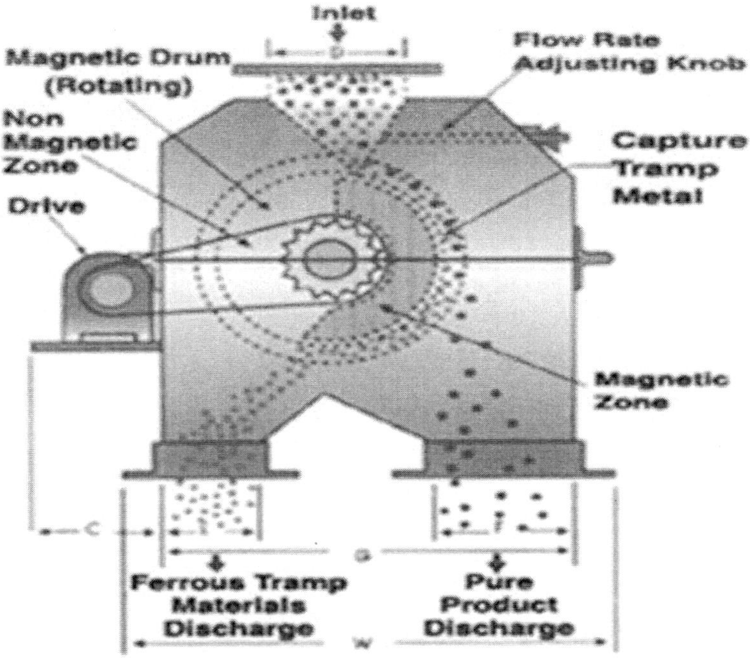

Magnetic separator

Rotary air screen cleaner

Screens

Screens are essential components of most cleaning machines. They come in different types, including perforated metal screens, wire mesh screens and nylon screens. Perforated metal screens are usually made of steel and have holes of different sizes. Wire mesh screens are usually made of stainless steel and have mesh sizes ranging from coarse to fine. Nylon screens are usually used for fine cleaning and are less durable than metal screens.

Types of Screens

- **Vibratory:** Vibratory screens are agitated by an eccentric unit and materials are separated by vibration of screens. These type of screens consist of perforated metal or mesh surfaces that separate impurities based on size. Most vibrating screens are inclined downward from the feed end. Vibration is provided to screen assembly only whereas the body and other surroundings are free from Vibration. The vibration of screen is responsible for performance of the following functions:

  i) It helps in providing passage to particles through opening of the screen.

  ii) It restricts clogging of the screen y particles that become trapped in the opening.

  iii) Because of vibration the particles are stratified over the screen surface and each particle has a chance to meet the screen opening

  iv) A continuous flow of particles along screen is possible.

- **Rotary screen:** This type of screen is either circular or rectangular decked. They possess the circular motion and affect the sifting action. They are very much efficient in separation of very fine particles. However, the capacity of rotary screen is very limited.

Rotary screen can be classified into two categories:

i) **Gyratory screen:** It is single decked screen which has horizontal plane motion which is circular at feed end and reciprocating at the discharge end.

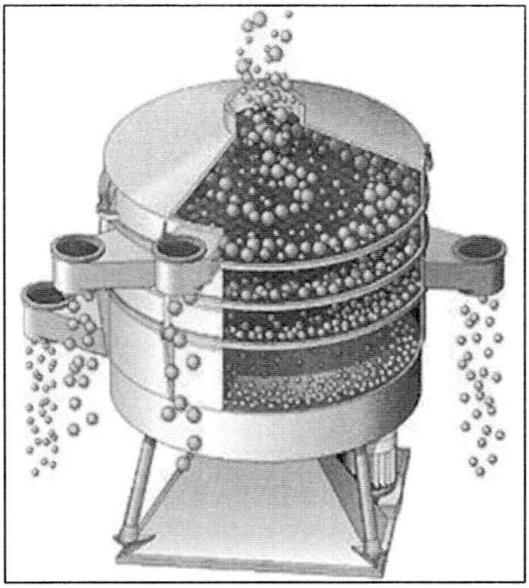

Gyratory screen

ii) Circular screens: Its motion in horizontal plane is circular over the entire surface.

- **Horizontal screen:** The pattern of vibration for the horizontal screen is back and forth while the inclined vibrating screen is circular.
- **Shaking screen:** In this type of screen material moves over rectangular surface down on an inclined plane. Motion of the screen is back and forth in a straight line. Unlike the vibrating screen, the shaker does not tumle or turn material en route except that some shaking screens have a step off between surfaces having different size openings, so that there may be two or three tumles over the full length of screen.
- **Revolving screen:** It is a cylinder that rotates about its longitudinal axis. The axis of cylinder is inclined along with the feed end to the discharge end. Sizing is achieved by having smallest opening screen at the feed end with progressively larger opening screens towards the discharge end.

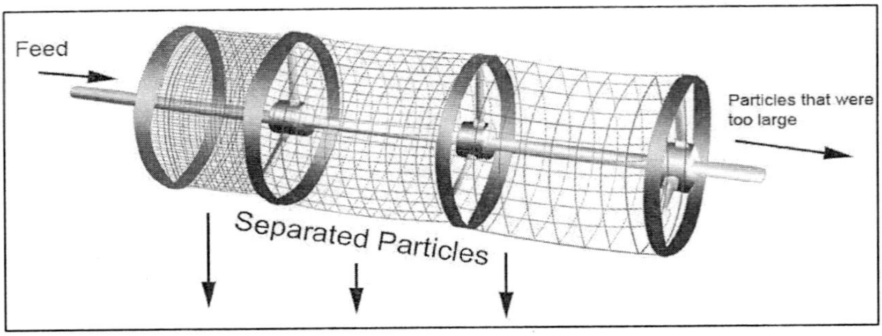

Rolling screen

Cleaning efficiency and quality control

The efficiency of the cleaning process can be influenced by several factors:

Cleaning parameters: Optimal cleaning conditions such as water temperature, pressure, and detergent concentration should be determined based on the produce type, degree of contamination, and desired cleanliness standards.

Equipment calibration and maintenance: Regular calibration and maintenance of cleaning equipment ensure consistent and effective cleaning. Proper functioning of brushes, screens, and sprayers is crucial for optimal cleaning efficiency.

Monitoring and testing: Regular monitoring and testing of cleaned produce for quality indicators, such as microbial load, pesticide residues, and visual appearance, help assess the effectiveness of the cleaning

Washing of fruits and vegetables

Fresh fruits and vegetables are requiring various amount of cleaning and washing before sending to market. Cleaning procedure is necessary which typically involves the removal of soil dust, adhering debris, insects and spray residues. Cleaning and washing also help in making the vegetables/fruits fresh and also improves appearance, it also helps in extending the shelf life of the produce. For cleaning of some fruit type vegetables (melons, brinjals, tomatoes, cucumber) they should be wiped with damp cloth. Many vegetables need trimming, cutting and removal of unsightly leaves or other vegetative parts. However, some type of produce can be washed, brushed or cleaned with a soft cloth. Washing is required to clean produce which has acquired latex stains from injuries from bad harvesting or handling, such as mango, papaya and most of the root crops. It is important to note that washing should be carried out only when absolutely essential and with only clean water. The washing of produce in recirculated water should be avoided because it can quickly become heavily contaminated with decay organisms, leading to rotting of the washed produce. Hypo chlorites or chlorine may be added to washing water but its use is not recommended for small-scale operations, because they are quickly inactivated by the organic material present in the dirty water. Washed produced to be treated with fungicide should first be drained after washing to reduce the danger of reducing the concentration of the fungicide When washing is not followed by chemical treatment, the washed produce should be spread out in a single layer on racks of mesh, in the shade but exposed to good ventilation to aid rapid drying.

Importance of washing

Washing fruits and vegetables serves several purposes:

- **Removal of dirt and debris:** Washing eliminates visible dirt, soil, and other foreign materials adhering to the surface of produce, making them more visually appealing.
- **Elimination of pesticide residues:** Many fruits and vegetables are treated with pesticides during cultivation. Washing helps reduce pesticide residues, promoting food safety.
- **Microbial reduction:** Fruits and vegetables can harbor various microorganisms, including bacteria, fungi, and viruses. Washing aids in reducing microbial contamination, thereby minimizing the risk of foodborne illnesses.
- **Extension of shelf life:** By removing surface contaminants, washing helps delay the decay process, maintaining the freshness and extending the shelf life of produce.

Cleaning methods for fruits and vegetables

- **Water washing:** Water washing is one of the most common and effective methods for cleaning fruits and vegetables. It involves immersing the produce in water or applying water sprays to remove dirt and contaminants. The water used should be potable, and the produce should be thoroughly rinsed to ensure the removal of all debris.
- **Brushing:** Brushing is particularly useful for fruits and vegetables with rough surfaces, such as root crops and certain melons. Soft or medium-bristle brushes are used to gently scrub the surface, dislodging dirt and debris. Care should be taken to avoid damaging the skin or causing bruises.

- **Detergent or sanitizer application:** Mild detergents or sanitizers can be added to the water during washing to enhance cleaning effectiveness. These solutions help remove surface contaminants, including waxes, oils, and microbial contaminants. However, it is essential to use food-grade and approved detergents or sanitizers and follow recommended concentrations to avoid any potential health risks.
- **High-Pressure water jets:** High-pressure water jets are effective for cleaning hardy fruits and vegetables with relatively smooth surfaces, such as apples, cucumbers, and bell peppers. The forceful water jets dislodge dirt and contaminants, providing thorough cleaning.
- **Steam cleaning:** Steam cleaning is a suitable method for delicate fruits and vegetables, as it effectively removes surface contaminants without excessive water absorption. Steam can be used in combination with brushes or as a direct application to the produce surface.
- **Ultrasonic cleaning:** Ultrasonic cleaning involves immersing the produce in a tank of water and subjecting it to high-frequency sound waves. The waves create microscopic bubbles that implode, generating tiny shock waves that dislodge dirt and contaminants from the produce surface.
- **Chemical-free cleaning methods:** In cases where chemical residues need to be minimized, chemical-free cleaning methods such as ozonation or ultraviolet (UV) light treatment can be employed. Ozone gas or UV light effectively kills bacteria and fungi, reducing the microbial load on the produce surface.

Machinery for cleaning of fruits and vegetables

- **Fruit and vegetable cleaning machine:** It is most common machine used for most of the fruits and vegetable cleaning. Generally, it clean 20 kg of fruit or vegetable at a time. It removes field soil, dust and surface microorganisms etc.
- **Multifunctional cleaning machine:** The machine is a multifunctional fruit and vegetable cleaning equipment with bubbles, spraying and brush available in cleaning with features such as highly cleanness, energy-saving, water-saving, stable and reliable. This equipment is suitable for cleaning fruits and vegetables. It is easy to operate, convenient in maintenance and wiring.

Fruit and Vegetable **Multifunctional cleaning machine**

Cleaning machine

- **Fruit cleaning machine:** The equipment is suitable for cleaning of ball-shape or oval-shape fruits and vegetables. The fruits and vegetables rotate continuously in all directions randomly. Brushing and spraying is in effect at a same time, with features such as high cleanliness. The machine is easy to operate, convenient in maintenance and wiring. (Fruits and cleaning)
- **Brush Type Vegetable & Fruit Cleaning Machine:** The fruit and vegetable raw materials are making irregular rotation under the effect of rotary brush roller to carry out spraying and brushing simultaneously. The brush is made of high temperature resistant nylon wire through two kinds of technologies such as hair planting and stainless-steel winding.

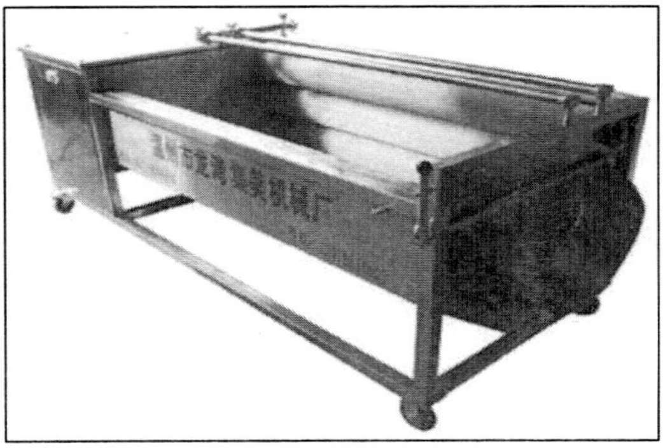

Fruit cleaning machine

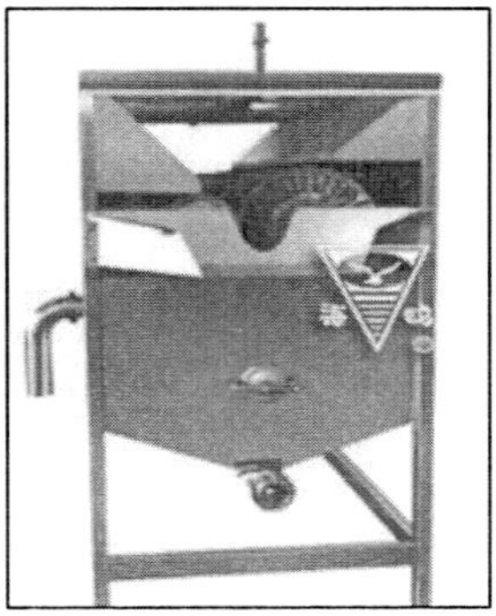

Brush Type Vegetable & Fruit Cleaning machine

- **Surf type fruit cleaning machine:** The equipment is mainly composed of water cabinet, material turning device, fan and lifter etc. It is widely used for soft washing of fruit and vegetable raw materials. The lifter can be made of complete stainless steel and engineering plastic. It can be additionally provided with spray cleaning device.

- **Roller with brush cleaning machine:** The fruit washing equipment consists of a roller with brush washing (cleaning) machine for washing fruits and vegetables. Roller with brush washing machine is made up of stainless-steel tube and brush. The brush is made of polyethylene, and will make revolution as the movement of stainless-steel chain. Fruits are driven to circumvolve and washed by brush. At the same time, the bad or rejected fruits are picked up by manual and then sent away by scrap conveying device.

Surf type fruit cleaning machine

Roller with brush cleaning machine

- **Rolling drum brush washing machine:** The fruit washing equipment consists of a rolling drum brush washing (cleaning) machine for washing root vegetables. Rolling drum brush washing machine is made up of electric motor, water pump, roller drum, supporter, riding wheel, brush, water spraying tubes, feeding funnel, cover board, water box, transmission shaft, supporter for motor, electric control switches, and other parts. Roller drum is driven to rotate by chain wheel of electric motor through stainless steel chain. When materials enter into the rotating roller drum, they are washed by spraying water and brush. There are two water boxes in our rolling drum brush washing machine. One is used to hold fresh water, and the other with filtering net in it is to recycle water.

Rolling drum brush washing machine

- **Cleaning efficiency:** The efficiency of the cleaning process depends on several factors, including the type of cleaning machine, the type of produce being cleaned, and the level of contamination. Some machines are more effective at removing certain types of impurities than others. For instance, air cleaners are better at removing dust and debris, while washers are better at removing dirt and contaminants from the surface of the produce.

Care and maintenance

Proper care and maintenance of cleaning machines are essential to ensure their longevity and efficiency. Regular cleaning of screens and other components is essential to prevent clogging and maintain their effectiveness. Regular inspection of machines for wear and tear and timely replacement of worn-out parts is also critical. Proper lubrication of moving parts is necessary to prevent rust and ensure smooth operation.

References

1. http://eagri.org/eagri50/AENG252/pdf/lec07.pdf
2. Gunathilake, D. M. C. C., Tiwari, A. K. and Kahandawala, K. T. (2018). Efficacy of washing treatment for extending the post-harvest shelf-life of papaya (Carica papaya). Int. J. Chem. Stud, 6, 2173-2177.
3. Madan, H.S. and Ullasa, B.A. 1993. Post-harvest losses in fruits. Advances in Horticulture. (Ed.) Chadha, K.L, and Pareek, O.P. Malhotra Publishing House, New Delhi. 4(4) p: 1795
4. Sudheer, K. P. and Indira, V. (2007). Post-harvest technology of horticultural crops (Vol. 7). New India Publishing.

Chapter - 7

Sorting and Grading: Sorting, Grading, Methods of Grading; Grading-Size Grading, Color Grading, Specific Gravity Grading; Screening, Equipment for Grading of Fruits and Vegetables, Grading Efficiency, Care and Maintenance

Authors

Pooja Bhatt
Assistant Professor, Chandigarh College of Hospitality, CGC, Landran, Punjab, India

Nivedhika Rai Vaid
Assistant Professor, Chandigarh College of Hospitality, CGC, Landran, Punjab, India

Vaishnavi
Ph.D. Scholar, Department of Food Science, Nutrition and Technology, CSK Himachal Pradesh Agriculture University, Palampur, Himachal Pradesh, India

Mamta
Ph.D. Scholar, Department of Food Science, Nutrition and Technology, CSK Himachal Pradesh Agriculture University, Palampur, Himachal Pradesh, India

Chapter - 7

Sorting and Grading: Sorting, Grading, Methods of Grading; Grading-Size Grading, Color Grading, Specific Gravity Grading; Screening, Equipment for Grading of Fruits and Vegetables, Grading Efficiency, Care and Maintenance

Pooja Bhatt, Nivedhika Rai Vaid, Vaishnavi and Mamta

Abstract

Agriculture is one of the major sectors and a main supplier to the financial growth of the country. Although sorting and grading are technically separate activities, the phrases are typically used synonymously in the food processing sector. Efficiency in manual sorting is entirely dependent on the operators, making it a very time-consuming, expensive, and ineffective procedure. Fruits and vegetables must be of very high quality while being exported to markets. Grading is crucial for fruits and vegetables since it increases their market value. Grading and sorting are done based on the size, color, volume, shape and texture. Human actions could be unpredictable, inefficient, and slow. According to the World Trade Organisation (WTO), modern marketing trends demand high-quality products. To ease labour shortages, save time, and enhance the quality of graded goods, farmers look forward to having a sufficient agricultural product grading equipment. Fruit grading is a crucial activity since it increases the grower's income and enhances the system's management, packaging, and general growth. This chapter will give insight to advance grading and sorting methods

Keywords: Grading, sorting, grader, specific gravity, color grading

Introduction

In India's socioeconomic development, agriculture is crucial. Various Agricultural commodities are produced all year long. Pre-harvesting and post-harvesting are the two phases. Fruits are sorted and graded as part of the post-harvest process. Grading of agricultural produce is primarily done manually in India. In the typical scenario of today, fruit and vegetables are manually sorted and graded according to maturity level before transportation. This labor-intensive, time-consuming manual sorting by visual inspection has the drawback of inconsistent and inaccurate judgements made by various humans. As a result, there is a desire for inexpensive cameras, and the availability of processing power makes it possible to employ machine vision-based systems to analyse this issue. The export market demands high quality graded agricultural produce.

One of the most important processes in food industry is sorting and grading (Arjenaki *et al.*, 2013). Sorting is the process of separating the healthy and clean fruits and vegetables from the ones that are rotten, damaged, or sick. Grading also refers to the process of classifying fruits and vegetables according to their colour, size, maturity stage, or level of ripeness. The two procedures are essential for preserving the quality and post-harvest shelf life of agricultural products. Grading also makes it easier for workers to group fruits and vegetables according to set standard criteria. Grading fruits according to their colour or state of ripeness, for example, would assist remove overripe fruits that easily produce ethylene to speed up the ripening process for the entire batch (Saltveit *et al.,* 1999). Commercial tomato growers typically employ sophisticated systems that call for exact criteria for sorting and grading their produce. However, the produce must still be sorted and sized to some extent before being sold or processed. Small-scale producers and retailers in underdeveloped nations, in contrast, may not follow grading properly.

Sorting

Similar to grading, sorting makes it easier to perform later processing steps like peeling, pitting, blanching, slicing, and container filling. It is useful for mass and heat transfer processes where the processing time depends on the size of the product (e.g., mass diffusion, heat conduction). Almost all food products go through some kind of sorting process. The necessity for sorted units in weight filling procedures, as well as the aesthetic and marketing advantages of offering uniformly sized or uniformly coloured units, are only a few of the perks. Additionally, sorted food units make it much easier to regulate procedures like sterilisation, dehydration, or freezing. They are also better suited to automated procedures like size reduction, pitting, or peeling (Grandison 2011).

Since weight is independent of the products' geometry, weight sorting is typically the most accurate procedure. Using spring-loaded, strain gauge, or electronic weighing devices built into conveying systems, eggs, fruit, or vegetables can be divided into weight groups according to their density. The heavier items are removed first, followed by the next weight category, and so on, using a sequence of tipping or compressed air blowing devices that are designed to activate at increasingly lower weights. These computer-controlled devices can also provide information on the distribution of sizes and amounts from various producers. The "catapult" principle, which uses spring-loaded catapult arms to launch units into various collecting chutes based on their weight, is an alternate technique. The comparatively low efficiency of weight sorting is a drawback.

Grading

Grading is the process of separating food products based on quality. The term "quality" refers to varied circumstances for different agricultural commodities as well as the relative significance of the basic material attributes. In actuality, grading involves a fair evaluation of all a material's characteristics, which has an impact on whether it is accepted as a food product and a working material for the processor. Grading sometimes entails the simultaneous examination of several traits, which might complicate mechanical grading and favour manual grading. The separation is based on the food product's size, colour, and form, which are designed to provide various characteristics (Omre and Saxena 2003).

Method of grading

Grading methods can be divided into two categories

1) Quality control process wherein samples statistically selected from a batch of food are used to determine the quality of the food through laboratory testing.
2) Process whereby the total amount of food is physically separated into different quality groups. This grading can be done manually or using advanced machinery.

For each product in the product category, grading is carried out using precise machinery. Foods are evaluated and sorted during food grading, which is frequently done manually. Food grading equipment is also employed, and this could involve grading products according to size, shape, and quality (George *et al.,* 2011). For the selling of fruits and vegetables, grading of quantity and quality is crucial. Sizers can be broken down into weight and diameter. Machines differ based on the item, grade, design, and size. By weight, this produces more than machine grade. Fruits are sorted based on quality using roller feed conveyors. It is built using the criteria listed in the table below:

Table 1: Criterion for Grading Agricultural Products

S. No.	Group	Standards
1.	Physical attributes	Shape, weight and size
2.	Technological attribute	Processing suitability or compatibility to existing equipment
3.	Organoleptic attributes	Aroma, taste, texture, color, ripeness or freshness
4.	Commercial attributes	Tradition, attractiveness, utility, variety and price

Benefits of grading techniques

1) Graded goods sell for a fair price.
2) Between buyers and growers, grading assists increase trust.
3) Increase marketing effectiveness by purchasing and selling a product you don't want.
4) Grading allows you to save the significant marketing expense of shipping and packing.
5) Increasing the earnings of distributors.
6) Grading enhances homogeneity in a product of a specific grade and serves as the basis for price.
7) Graded products are less negotiable, hence it will give more profit to the producer.
8) Grading will set a solid standard of comparable grades and materials competitiveness, which will help to develop perfect competition in the market.

Size grading

Size grading is the upmost parameter of grading but in India most of the size grading done manually. Size is also important determinant for amount of wastage that occurs during processing and also size grading help in easier handling of huge quantities. A variety of screens are used to separate agricultural products by size. A fixed aperture screen employing vibratory, gyratory, or stationary screens and a variable aperture screen utilising a belt, scew, cable, or roller are the two categories into which the screen designs most frequently employed in the food industry can be divided. Cereals, herbs, and nuts are sized using multi-deck screens, whereas turnips, potatoes, and potatoes are sized using fixed aperture flat-bed screens. Fruit and vegetable size grading uses variable aperture screens with continuously variable aperture of the roller, screw, and belt types (Sivasankar 2019).

Color grading: In grading, color is the most important parameter in defining the quality of any agricultural product. With the quality of agricultural product, color grading also give nutrient content of the product as bright in colour indicates high content of nutrients (Bhatt *et al.,* 2023). Color is often a measure of maturity, presence of defects, or the degree of processing (Fig 1). Many times consumer decision to accept or reject a particular product is depends upon its color. Color grading is a crucial step in the food processing that directly affects profitability, because the quality of agricultural products is often associated with their color (Satpute and Jagdale 2016). The majority of currently used automatic colour grading systems either use a set of colour separation characteristics, frequently in three-dimensional colour spaces, or directly compare product colour against a preset and fixed set of reference colours to determine colour quality. The user cannot easily change colour preferences or grading factors using these techniques. (Jye *et al.,* 2011).

Noordam and Otten (2000) conducted research on high-speed colour vision-based potato grading and quality checking. For the purpose of grading and inspecting the quality of potatoes, a fast machine vision system has been created. The vision system assigned grades to potatoes based on their size, shape, and outward flaws like mechanical damage, rhizoctonia, common and silver scabs, as well as growth cracks. The potatoes were scrutinised under the 3-CCD line-scan camera as they flew by. Due to the utilisation of mirrors and the absence of product holders, a complete view of the potato was gained.

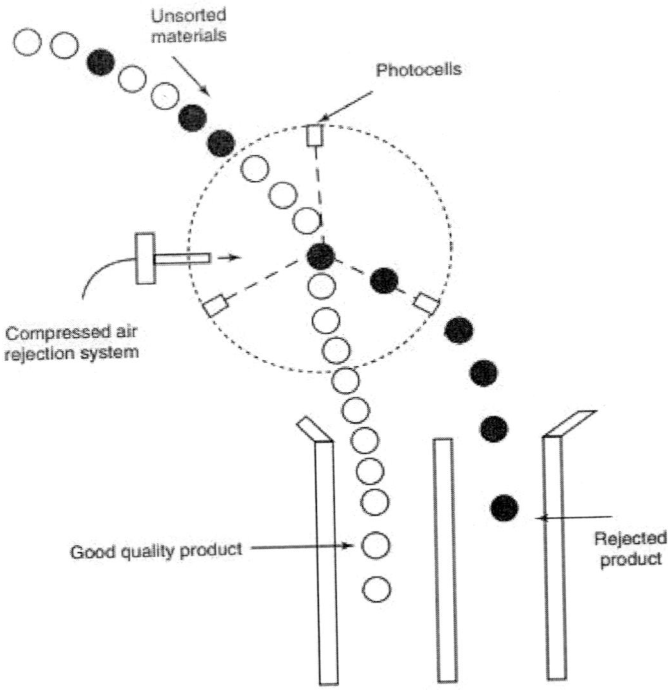

Fig 1: Color grading

Specific gravity grading

The internal quality of agricultural produce, such as internal breakdown, water distribution, decay, and insect infestation, is frequently related to specific gravity, an important indicator for the quality assessment of that produce (Teerachaichayut *et al.,* 2015). Fruits' specific gravity is determined by comparing them to water's density. Farmers and agriculturalists will typically measure the ripeness of the fruits by using floating approaches based on differences in density when using the specific gravity methodology. However, it takes a lot of time and is inconvenient. It is possible to estimate the fruit's maturity using the specific gravity measurement. Therefore, methods for processing images were suggested in order to calculate the specific gravity of fruit (Khalid *et al.,* 2017).

Screening

Many different fruits are sorted using vibrating screens composed of plastic, stainless steel, and copper that doesn't react chemically to the produce. The rotating screen or vibrating screen receives the material to be graded. The smallest substance near the screen's edge moves through the centre and then the largest material. In this way, it creates diverse classes of vegetables like tomato and potato as well as fruits like orange, apple, and kinnow. The material at the top of the screen can be recreated to produce new fractions (Badhe *et al.*, 2011).

Fig 2: Screening of potatoes

Equipment for grading of fruits and vegetables

Fruits are frequently graded manually in the nation. Manual grading is an expensive and time-consuming process, and the business is impacted because there aren't enough workers available during the busiest times (Narvankar and Jha, 2005). Graders were first created more than 50 years ago, and they were basically a rough slat with a hag connected to the end. On the slat, goods were examined before being manually placed into the bag. These were referred to as slat graders, which encouraged the creation of mechanical graders. Grading hasn't altered all that much in the past 50 years. However, the grading procedure is now entirely automated. A bag was attached to the end of a chain conveyor belt that served as a mechanical grader. Produce that was smaller fell through the cracks, simplifying the grading procedure.

Grading is still done by hand in India. There was a search for automated solutions due to a lack of consistency in the process and a labour scarcity. Vegetable grading provides a larger emphasis on quality assessment in order to respond to market demand, which increases the need for better, more precise grading and sorting procedures. Onions and potatoes, which vary in size, served as a base for categorising vegetables into other groups. Every nation that produces vegetables has developed its own criteria for various grades while taking the needs of the market into consideration (Londhe *et al.*, 2013)

Grading based on size includes divergent roller type graders, expanding pitch type graders, inclined vibrating plate graders, and counter rotating roller graders with an inclination. Agriculture goods' specific gravity and density are used to grade their weight. Quality assessment is given increased importance in order to be responsive to market demand, which increases the requirement for enhanced and more precise grading and sorting procedures.

1. Image processing

It is an electrically powered device. Fruit images are taken, and after that, grading is done using image processing. The three basic processes that make up the machine vision inspection system are a) picture acquisition, b) image processing, and c) decision-making. Image acquisition is done using cameras with the proper illumination. Analogue to digital conversion is done for the visual data. To extract the necessary object attributes and quality metrics, image processing hardware and software analyses the acquired images. Software makes a choice on the fruit quality based on these findings while taking the end user's needs into account (Kamble *et al.,* 2019).

2. On-farm rotary disc size grader

On-farm rotary disc size is a low-cost grader with different diameters used in the laboratory to grade spherical fruits and vegetables, and a low-cost, on-farm rotary disc size grader was created. The best disc speed for properly separating the various grades was found to be between 60 and 70 rpm. The grader will undergo testing for various fruits and vegetables in the field and in a lab. (Sravan and Tejaswini 2020).

3. Manual grading

Fruits, vegetables, eggs, and other food products are primarily graded manually. The grader classifies the food into groups based on quality and makes a fair judgement about the overall quality. Colour cards that exhibit different grades in relation to surface proportions indicating fruit colour, corrosion, surface flaws, etc. can be used to grade apples. Models of cherries constructed of plastic spheres with nylon stems are similar to cherries. By candle, the eggs are carefully graded. In this kind of test, an egg rotates in front of a bright light while its translucent shell rotates in a dark booth.

Manual grading has a few problems, including:

Manual Grading of fruits and vegetables is an exhausting task that takes a lot of work. Manual grading needs more labors and skill workers. In India, the majority of the fruit grading-including that of mangoes, oranges, kinnow, and vegetables-is done manually (Sivasankar 2019).

4. Roller grading

These graders are quick, accurate, and cause minimum fruit damage. In the fruit industry, these are frequently utilised. Each roller spins in the opposite direction of the clock. To provide each piece of fruit an opportunity to record its smallest measurement in the grader, the fruit is continuously rotated. For removing little fruit, twigs, and leaves, roller conveyers are employed with fixed space among the rolls (Sravan and Tejaswini 2020).

5. Weight grader

Based on the product's weight, the grading is finished. This method preserves fruits the least while being accurate and quick. Additionally, this is employed for large-sized products like oranges, apples, potatoes, mangoes, eggs, tomatoes, etc. These are specially designed for sorting materials that, because to their shape or texture, cannot be handled by other processes. The material to be graded is automatically fed into separate cups, where it is placed in the sorter where it is indexed with the aid of spring-loaded trips. A fruit grader that can classify fruits into one of four weight-based categories (A, B, C, and D). Grade A (> 200 g), Grade B (150–200 g), Grade C (100–150 g), and Grade D (100 g) are available. The performance of the grader is assessed at various speeds (5, 10, 15, 20, 25, and 30 rpm) (Omre and Saxena 2003).

A sapota fruit grader was created by Ukey and Unde (2010). The design and development of a sapota fruit grader based on the diverging roller type principle was done to boost the output of fruit grading while reducing time and labor costs. For the highest efficiency of 89.5%, the optimal combination of roller speed, its inclination, and roller gap was found to be 223 rpm, 4.50, and 38 to 64 mm, respectively. The machine had a 1,440 kilogramme per hour capacity and cost Rs. 11,450 (without an electric motor).

Grading efficiency

Grading efficiency depends upon the method choose for grading of fruits and vegetables. Grading efficiency is low in manual grading as compared to the machine grading (Sivasankar 2019). Computerised grading has logistic software which increases the efficiency of grading with fewer errors. For e.g. the diverging roller type apple grader did an effective job of classifying the apples according to their size and shape. The range of the grading efficiency was 57.9 to 84.2% (Muzamil *et al.*, 2018).

Care and maintenance

One of the most popular maintenance techniques is scheduled maintenance. In the machine grading, weekly check the oil level in the grader.

In the Screen grader, check for any plugging and clean daily. When small-size particles become wedged or stuck, they block the apertures and cause plugging or pegging. In general, plugging refers to loose particles or stones that become lodged in media holes, whereas pegging refers to particles that become jammed in openings. Also, there should be daily inspection of bolt joints for tightness of the parts of graders.

Conclusion

Grading and sorting of agricultural products such as fruits and vegetables are very important in post-harvesting process. Manual grading is costly, time-consuming, and ineffective. The primary criteria for grading and sorting are internal and external quality characteristics. Colour, size, volume, shape, and texture are examples of extrinsic elements. Among these, size and colour are the primary criteria for classifying fruits. Sorting apples, tomatoes, and other produce based on size is a simple and affordable process. Direct colour mapping is a flexible and effective method for color-based grading. Grading schemes provide us with a wide range of details, including size, shape, colour, fault, and interior quality which help in marketing. Grading and sorting based on other external factors cannot be done manually, and while grading and sorting based on size can be done manually, the results are not accurate. Therefore, fruit and vegetable quality inspection requires automation."

Reference

1. Arjenaki O.O, Moghaddam P.A, and Motlagh A.M, 2013. Online tomato sorting based on shape, maturity, size, and surface defects using machine vision. Turkish Journal of Agriculture and Forestry, 37(1), 62-68.

2. Badhe V.T., Singh P., and Bhatt Y.C., 2011. Development and evaluation of mango grader. Journal of Agricultural Engineering, 48(2), 43-47.

3. Bhatt P, Sharma S, Grover K, Sharma S, Dharni K, and Dhatt A.S. 2023. Optimization of Nutritional Composition, Bioactive Compounds and Antioxidant Activity in Broccoli (Brassica oleracea) Microgreen Sprinkler using Alternate Drying Techniques. Asian Journal of Dairy and Food Research.42,1-8.

4. Grandison, A. S., (2011) Postharvest Handling and Preparation of Foods for Processing. Food Processing Handbook.1–30.

5. Jye D., Lee, Archibald J., and Guangming X., (2011). Rapid Color Grading for Fruit Quality Evaluation Using Direct Color Mapping. Automation Science and Engineering, IEEE Transactions on. 8. 292 - 302.

6. Kamble A.B., Pandav Pornima T., Patil S., Salunkhe K.D., (2019) Grading and sorting of apple by using image processing. International Research Journal of Engineering and Technology. 6 (3), 7878-7883.

7. Khalid N.S., Abdullah A.H., Shukor S.A.A., Mansor F.S.A.S., H. and Dalila N.D.N., (2017) Non-Destructive Technique based on Specific Gravity for Post-harvest Mangifera Indica L. Cultivar Maturity. *Asia Modelling Symposium (AMS), Kota Kinabalu, Malaysia*, 113-117.

8. Muzamil M., Wani J., Mustafa A., Mohiuddin M., Mir A., Faisal S., and Nirbhavane S. (2018) A Low-Cost Divergent Type Apple Grader for Kashmiri Apples. International Journal of Current Microbiology and Applied Science. 7(4), 3561-3569.

9. Omre P. K., and. Saxena R.P., (2003) Design and development of multi-fruit grader. Journal of Agricultural Mechanization in Asia, Africa, and Latin America, 34 (3), 39 - 52.

10. Saltvei M.E., (1999) Effect of ethylene on quality of fresh fruits and vegetables. Postharvest Biology and Technology, 15(3), 279–292.

11. Saravacos, George D., Maroulis, Zacharias B. (2011).Food Process Engineering Operations. CRC Press. 198-199.

12. Sivasankar B., (2019) Food processing and preservation. Prentice Hall India Learning Private Limited. Ltd.174-178.

13. Teerachaichayut S., Yokswad W., Terdwongworakul A., Jannok P., and Fernandes S.V. (2015) Application of Image Analysis for Determination of Mangosteen Density. Journal of Advanced Agricultural Technologies. 2(2), 92-97.

Chapter - 8

Parboiling Process, Changes During Parboiling, Parboiling Methods, Advantages and Disadvantages of Parboiling with Respect to Milling, Nutrition, and Cooking Quality of Grain. Milling: Milling, Polishing, Grinding, Milling Equipment, Dehuskers, Polishers, (Abrasion, Friction, and Water Jet), Flour Milling Machines, Pulse Milling Machines, Grinders, Cutting Machines, Oil Expellers, Machine Efficiency, and Power Requirement

Authors

Neha Naijo Areekal
Ph.D. Scholar, Indian Institute of Technology, Tirupati, Andhra Pradesh, India

Ramachandran. R
M Tech Food Technology, College of Food and Dairy Technology, Koduveli, Tamil Nadu, India

M. Keerthana
Assistant professor, Karpagam Academy of Higher Education, Coimbatore, Tamil Nadu, India

Hemamalini
Assisstant professor, Sree Sastha Institute of Engineering and Technology, Chennai, Tamil Nadu, India

Chapter - 8

Parboiling Process, Changes During Parboiling, Parboiling Methods, Advantages and Disadvantages of Parboiling with Respect to Milling, Nutrition, and Cooking Quality of Grain. Milling: Milling, Polishing, Grinding, Milling Equipment, Dehuskers, Polishers, (Abrasion, Friction, and Water Jet), Flour Milling Machines, Pulse Milling Machines, Grinders, Cutting Machines, Oil Expellers, Machine Efficiency, and Power Requirement

Neha Naijo Areekal, Ramachandran. R, M. Keerthana and Hemamalini

Abstract

Around the world, 130 million tonnes of paddy are parboiled each year, and 3-4 million tonnes of valuable parboiled milled rice are transported in international trade. Parboiling originated and is largely practiced in India, Bangladesh, and Sri Lanka. The process of milling is an ancient technique and holds a history as old as ancient human civilization. With the revolution of human civilization, the development and modernization of the milling process came up. Thus, the grinding process was initiated with the invention of mortar & pestle to grinding rolls and evolved to the present modernized milling equipment like hammer mills, stone mills, etc. adopting similar principles with certain modifications. This chapter deals with the parboiling process, changes during parboiling, parboiling methods, and the advantages and disadvantages of parboiling with respect to milling, nutrition, and cooking quality of grain. milling: milling, polishing, grinding, milling equipment, dehuskers, polishers, (abrasion, friction, and water jet), flour milling machines, pulse milling machines, grinders, cutting machines, oil expellers, machine efficiency, and power requirement.

Keywords: Parboiling, milling, grinders, cutters, machine efficiency

Introduction

Parboiling is a hydrothermal pre-milling procedure that involves soaking, steaming, and drying before gelatinizing the starch component of rice. Partial cooking for rice intact with husk is called parboiling, which involves soaking, steaming, drying, and milling. Parboiled rice takes less time to cook and is firm and less sticky. This procedure alters the grain's physical, chemical, and organoleptic properties, which have a significant impact on the grain's milling, storage, cooking, and eating qualities (Bhattacharya *et al.*, 2013).

Milling involves the process of dehulling, splitting (in case of pulses) (Wood and Malcolmson, 2021) and converting the larger size particles into smaller or finer particles. Depending upon the requirement of the end product and specific food commodity, the products of milling namely, flakes, broken, splits or powders are achieved by flaking or crushing or grinding/powdering processes. The evolution of ancient milling equipment is shown below in Fig 1.

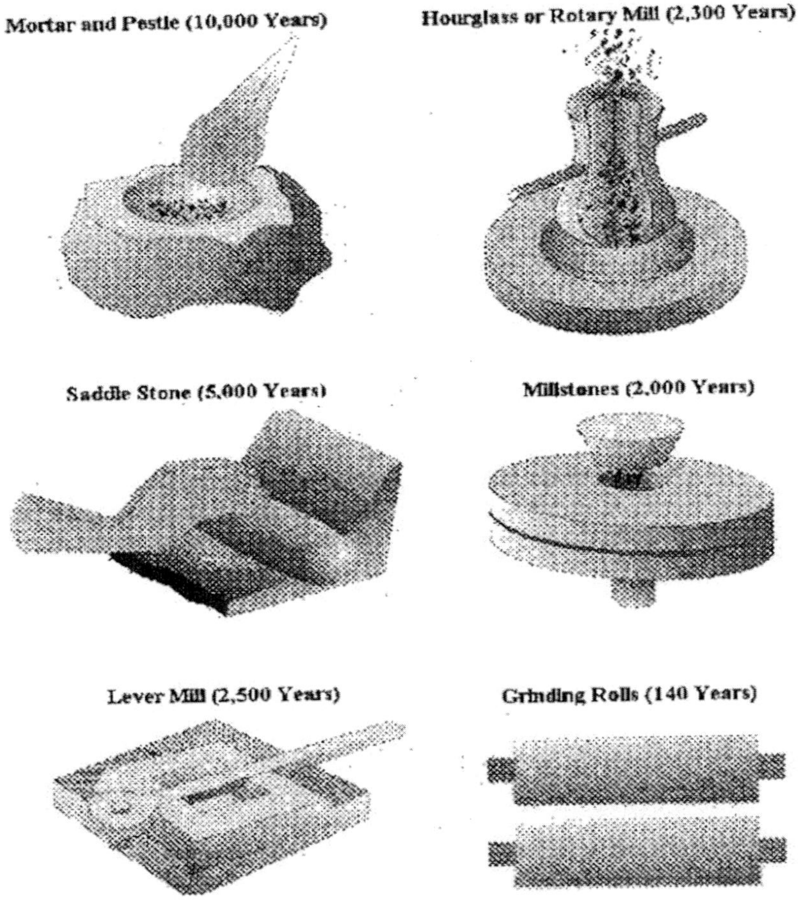

Fig 1: Evolution of ancient milling equipment (Chakraverty *et al.*, 2003)

Parboiling process

When rice is parboiled, its nutritional value and milling yields are increased because the rice is first soaked, steamed, and then dried. Amylose-lipid complexes are formed, proteins form disulfide bonds, and starch turns into gelatin during the parboiling process. The undesirable "white core" of ungelatinized starch in paddy rice is decreased to at least 30% moisture content during soaking (Bhattacharya, 2004).

It is difficult to ensure equitable distribution of water because the path, pattern, and rate of water penetration into a rice kernel can differ greatly depending on its morphological structure, such as the arrangement of starch granules in the endosperm and the degree of milling. Endosperm proteins in native and parboiled rice both act as inhibitors of water absorption (Derycke *et al.,* 2005).

The reduced ability to be extracted after parboiling shows that these proteins are denser because they inhibit the starch granules from expanding during heat operations. The substances amylose and amylose-lipid complex are also known to inhibit swelling and starch solubilization during cooking.

Therefore, the starch phase transition and the physical features of the grain will surely be hampered by the uneven distribution of water and low water content within rice kernels (Buggenhout *et al.,* 2013). The process flow diagram for making paddy parboiled rice includes steps like cleaning, soaking in warm water for 24 hours, steaming, drying, milling and dehusking, grading by quality, packaging, and storing.

Procedures and processes of parboiling paddy rice

The observed unit operations involved in paddy rice parboiling processes includes.

Sieving

If in the case of broken grains in the paddy, sifting is carried out. They are divided using wire netting to stop the broken kennels from heating and fusing the contents together during parboiling. The thick and sticky liquid from the cooked broken-kernel grains usually heaps the grains on top of one another and obstructs effective, rapid drying. Inattention could lead to the growth of fungi and the spoilage of the rice.

Washing, Floatation, and Sedimentation

To remove dirt and other particles, the paddy is immersed in water and rapidly stirred. While the larger pieces of trash sink, the lighter ones float. The paddy rice is scooped out, leaving the heavier materials (sand, stones, pieces of metal, etc.) at the bottom and being emptied away while the floating trash (dead insect parts, weeds, unfilled seeds, etc.) is skimmed off. Depending on how much dirt is in the rice, washing may be done twice or three times. According to science, washing the rice gets rid of the dirt, fungus toxicity, and other impurities.

Boiling

The partially cooked paddy is then placed in water on a hot pan and heated to a warm temperature of 35 to 40°C. As a result, more water may be absorbed by the paddy, and various enzymes' activities are reduced. Some microbes and the byproducts of their cells may be killed off or rendered inactive at this temperature.

Soaking

Paddy rice is soaked by being submerged in warm water over the course of a night (10–24 hours), allowing the rice to absorb the water and become moisture saturated. The basic scientific idea is the activation of enzymes. A conducive environment is created for the transformation of enzymes and pigments after cooling down. Toxins are diluted and microbiological illnesses may be eradicated or rendered dormant at such a high moisture level.

Steaming

After the paddy has soaked, it is steam-cooked in a container lined with a little water. The grain's high moisture content is used during the steaming process to gelatinize the starch. Jute sacks and polythene sheets are used to cover the pot, allowing steam to circulate throughout the entire object. The top helps the molecules in each gelatinous kernel flow inside by preventing the steam from quickly leaving and creating a partial pressure over the contents. When the paddy begins to split open its husks and steam vapour is rising all around the kettle, the steaming is complete. Heat also evaporates superfluous moisture from the grain, preserving concentrated nutrients in the kernel. Toxins and other hazardous substances.

Drying

Soon, the steaming paddy is spread out to dry in a wide open space. Too much sun exposure causes the kernel to dry unevenly, which leads to fracture during milling. The ability to exclude moisture and concentrate solutes in the kernels are two further scientific advantages of drying processes. On steamed paddy, drying also inhibits the development of microbiological pathogens. It makes dehusking (milling) easier by delocalizing the gelatinous amylase starch from the husk and compacting it into a bulk.

Milling

Soon, the steaming paddy is spread out to dry in a wide open space. Too much sun exposure causes the kernel to dry unevenly, which leads to fracture during milling. The ability to exclude moisture and concentrate solutes in the kernels are two further scientific advantages of drying processes. On steamed paddy, drying also inhibits the development of microbiological pathogens. It makes dehusking (milling) easier by delocalizing the gelatinous amylase starch from the husk and compacting it into a bulk.

Winnowing

Winnowing can be carried out manually or automatically. Winnowing the rice removes the husks and other impurities before it is manually pounded with a mortar and pestle. Winnowing is a method for getting rid of impurities like dead insects and dirt from rice.

Sorting/Grading

Commercial parboiling businesses go one step further and hand-pick any rice that is discoloured or has foreign pollutants before dividing into classes. Winnowing can be done automatically or by hand. When paddy rice is physically ground using a mortar and pestle, the husks and other impurities are removed by winnowing the rice. By winnowing, rice can be cleaned of dirt, dead insects, and other contaminants.

Changes during parboiling

Biochemical changes during parboiling

- The enzymes amylase, protease, phosphatase, and beta-glucosidase in rough rice (paddy) were activated during the soaking step of parboiling.
- Parboiled brown rice contained considerably more reducing sugar but less sucrose and free amino acids.
- On milling, there was a considerable loss of sucrose and free amino acid from raw rice, but very little from parboiled rice.
- A greater part of the sugar leached out into the water during soaking.
- In India and Pakistan, it is noted for its good storage properties and special taste.
- In terms of food preparation, processing, or preservation, parboiling is a crucial technique.
- The process of parboiling is likely to drive nutrients, especially thiamine, from the bran to the endosperm.

- Hence parboiled white rice is 80% nutritionally similar to brown rice during the process of gelatinization, alpha amylose molecules leach out of the micellar network and diffuse into the surrounding aqueous medium outside the granules.
- The granules become fully hydrated, producing a maximum in the measured viscosity.

Chemical changes

- The water-soluble substances (vitamins and minerals) are dissolved and allowed to permeate the grains.
- The endosperm's lipoid components dissolved.
- The gelatinized starch appears like a compact and homogeneous mass.
- The lipoid matter is separated and goes deeper into the compact mass of gelatinized starch.
- The germ and the outer portion of the endosperm contain fat-soluble compounds that are dissolved and distributed throughout the grain.

Physical changes

- There are fewer broken grains, which improves the milling yield and quality.
- Even without milling, parboiled rice can be stored more effectively and for a longer period since germination cannot occur and the endosperm's compact texture makes it more resistant to insect attacks and doesn't allow it to absorb moisture from its surroundings.
- Parboiled rice has a longer shelf life and is less likely to go rancid.
- The water content of the grain can be reduced through drying to a level that is ideal for milling.

Sensory changes

- Due to its texture and firm consistency, cooked parboiled rice has a higher digestibility ratio.
- After cooking, the grains are firmer and tend to be less sticky.

Parboiling methods

Methods of parboiling and effect of parboiling on milling quality

The method of parboiling may be classified as follows;

S. No.	Traditional Methods	Modern Methods	Methods under investigation
1.	Single boiling	CFTRI method	Brine solution method
2.	Double boiling	Jadavpur method	Kisan continuous method
3.		Avorio process	Pressure parboiling method
4.		Converted process	RPEC method
5.		Malek process	Sodium chromate method
6.		Cristallow process	Parboiling with heated sand method
7.		Fernandes process	
8.		Schule process	
9.		Rice growers association of California process	

Traditional method: The conventional method entails soaking the paddy for 24 to 48 hours or longer in water at room temperature, steaming it in kettles under pressure, and drying it in the sun. Paddy is soaked in regular water for 24-72 hours and then steam-cooked in a single boiling process. When using the double boiling method, the row paddy is first injected with steam in the steaming kettle before being soaked. Hot paddy increases the soaking water's temperature to 45-50°C, which helps to shorten the soaking period to 24 hours.

Modern method

CFTRI method: In this process, clean water is poured into parboiling tanks, which are then heated to a temperature of about 85°C by steam being forced via coils inside the tank. In parboiling tanks, hot water is occasionally pumped from other sources. The tank's finished paddy water mixture maintains a constant temperature of roughly 70°C. Water is discharged from the paddy after soaking it for 3 to 3.5 hours. The water discharge value is left open to allow condensed water to be removed while steaming. Through the open steam coils, the soggy paddy is exposed to steam at a pressure of approximately kg/cm^2. Paddy is soaked and steamed in the same manner. By opening the bottom door, the parboiled paddy is removed and dried either by the sun or a mechanical dryer.

Jadaopur University method: All the operations of this method are fully automatic and the average processing time is five to six hours. Paddy can be thoroughly soaked in water that is between 60 and 70 degrees Celsius for 1 to 3 hours, while the maximum amount of time for steaming is 3 minutes. After steaming and drying, the paddy is rapidly cooled. Drying takes place in a rotary steam-jacketed high-temperature Air dryer. Two different approaches may be used in this process. In the first, soaking and steaming occur in the same tank, whereas in the second, these two actions are carried out separately in a horizontal apparatus. Both times, saturated steam is employed. The steaming paddy quickly cools in a cold air drought.

Pressure parboiling method: This parboiling technique was created in Tiruvarur, Tamil Nadu. The parboiling process involves forcing water vapor under pressure into the paddy to introduce moisture. The starch in the kernel gelatinizes as a result. At 85 to 90 °C, the paddy is soaked for 40 minutes. It is then steam-cooked under pressure for 18 minutes. Air that has been trapped is forced out by water vapor that enters the kernel.

According to reports, the entire procedure takes between one and five hours. This technique yields rice that is uniformly coloured and almost yellowish. The main benefit of this method is that the paddy soaks for less time.

Avorio process: Paddy is stored in a perforated basket and transported into a hot water tank for soaking in the Italian Avorio process. It is then steamed under pressure in a revolving cylinder and dried in hot air dryers.

Conversion process: The process was developed in the USA. This paddy is immersed in cold water, the air is drawn out of it in a vacuum, it is then steamed under pressure, and it is dried in a vacuum dryer.

Malek process: Another technique was created in the USA and involves soaking the paddy at 100°F for 3 to 6 hours, steaming it for 15 minutes, and then drying it in a low-temperature hot air dryer.

Fernandez process: The procedure, which is widely utilized in Latin America, is comparable to hot soaking techniques.

Advantages and disadvantages of parboiling with respect to milling

S. No.	Advantages	Disadvantages
1.	Simple to dehusk, and fewer milling losses	Extra capital investment
2.	Higher yield of head rice and retains more protein.	Reduce polishers capacity
3.	More resistant to fungus and insect infestation.	Attacked by pores and require more energy
4.	Nutritional value of rice increases and Shelling is easier.	Aids to cost of drying
5.	Water soluble B vitamins, thiamine, riboflavin and niacin are higher	Dark colour because of extended soaking
6.	Does not become sticky mass	More fuel and longer time
7.	Milling yield and Quality increases	More oil and rancidity

8.	Milled rice turns into translucent and brilliant	Longer time to prepare
9.	Longer shelf life	Natural antioxidants are destroyed
10.	Reduces Breakage rate and improved yield	Mycotoxins can occur

Nutrition, and cooking quality of grain

Effect of parboiling on milling quality

Several variables contribute to kernel breakage in row rice milling. One of the main causes of breaking is kernel cracking. Delays in harvesting, threshing, or quick-drying caused cracks to form. Machinery that is old and chalky affects the output and quality of milling. Rice breakage is influenced by milling conditions, namely by the temperature, relative humidity, and amount of milling.

Breakage happens during shelling or husking operations. Due to the gelatinization of the starch during the parboiling process, the cracks, incomplete grains, and chalkiness in the paddy are entirely cured.

The rise in head production of rice during polishing, the polish percentage, and the reduction in breakage over time are the most advantageous aspects of parboiling, however, parboiled rice needs more time than row rice to polish to the same degree. For the same level of polishing, parboiled rice needs three to four times as much abrasive load as row rice.

Parboiled rice requires less polishing than row rice, depending on the consumer's preference for the color of the rice. For instance, if the consumer wants to remove 80% of the bran from their rice, parboiled rice only requires 3% polishing, whereas row rice requires 4% polishing.

Milling quality

The degree to which the endosperm is still intact after milling is referred to as milling quality. Total milling yield, also known as head rice yield (milled rice kernels with an endosperm length of at least three quarters), is a measure of the quality of the milling process. The procedure of milling should not result in breakage.

Nutritional quality

Compared to milled rice, brown rice has more nutrients (minerals and vitamins). However, the genetic diversity of rice grown around the world affects the status of nutrients.

Physiochemical and nutritional changes during parboiling treatment

Cereal grains' endosperm is primarily made up of polygonal starch granules. Air and moisture are present in the interparticle spaces, also known as voids. Due to this, it appears opaque. Additionally, during ripening, the grain develops fissures and breaks that cause the rice to break when it is milled. The endosperm of cereal grains contains mainly polygonal starch granules. The voids or interparticle spaces are filled with air and moisture. That's why it looks opaque. There are also cracks and fissures in the grain that occur during ripening that cause the rice to crack during milling.

The two biggest alterations that take place during parboiling are the dissolution of endosperm protein structures and starch gelatinization. Proteins and carbohydrates swell and fill the interior air layer. The endosperm's fissures and crevices are sealed, making the grains transparent and sturdy and reducing grain breakage during milling.

The color of harvested rice depends on many factors, including rice variety, soaking time and temperature, steaming time and temperature (vapor pressure), and drying time and temperature.

Rice type, soaking time and temperature, steaming time and temperature (vapour pressure), and drying time and temperature are only a few of the variables that affect the colour of harvested rice. Rice that has been parboiled takes longer to become as soft as rice of the same type that has not been cooked. Protein and starch losses in rice that has been parboiled are very small. Parboiled rice loses fewer nutrients after milling because water-soluble nutrients, like water-soluble B vitamins, diffuse into the endosperm during parboiling.

In parboiled rice, where modest dextrinization and lipase degradation take place during parboiling, vitamin E content is particularly noticeable. Some of the natural antioxidants that are destroyed by heat during parboiling can increase the rancidity of parboiled rice when it is stored.

Effect on cooking quality

Compared to uncooked rice, parboiled rice takes longer to cook. However, it is looser, less sticky, and releases fewer particles into the cooking water (Champagne *et al.*, 2004). Sticky, and releases fewer particles into the cooking water (Champagne *et al., 2004*). It also holds its shape better. Even without being steamed, the water absorption rate will decrease if it is submerged at 70°C or higher. When the soaked rice is heated or steam-cooked, the water absorption steadily reduces as the heating intensity rises, reducing solids loss while doing so (Dendy *et al.*, 2001).

How much the cooking quality is impacted depends on the combined heat treatment during soaking and streaming. Parboiled rice has a tougher texture than cooked uncooked rice in addition to boiling more slowly (Juliano *et al.*, 1985).

Even though it has a distinct color after being parboiled, the rice is almost as white as raw rice once it has been cooked. Parboiled rice expands less in length but more in breadth during cooking compared with raw rice. For this reason, parboiled rice appears more stocky and round after cooking (Pillaiyar, 1988).

Milling

Axially asymmetric pieces with several characteristics, including holes, slots, pockets, and even three-dimensional surface contours, are frequently produced by milling. Completely milled parts frequently include components like fasteners or brackets that are made specifically for the part and are used in small quantities, potentially for prototypes.

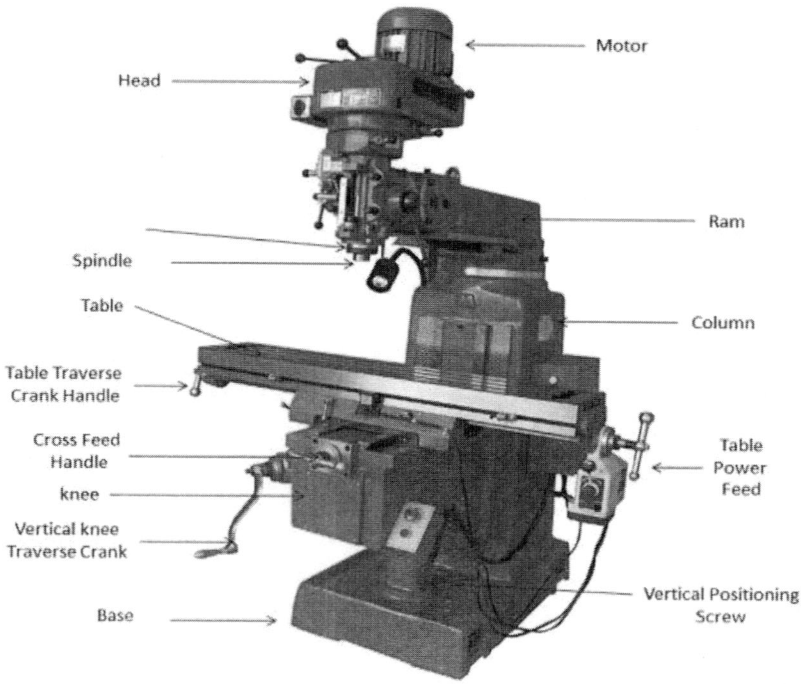

Fig 2: Milling equipment

The creation of tools for other operations is another use for milling. For instance, milling is commonly used to create three-dimensional molds. In order to add or improve characteristics to parts that were made using a different process, milling is frequently employed as a secondary process. An item whose basic shape has already been produced is perfect for adding precision features since milling offers excellent tolerances and surface finishes (Figure 2).

Grinding

Grinding is a chip-junking process that uses an individual abrasive grain as a slice tool. Grain and single-point slice tool conduct differ significantly in a number of ways, including the irregular forms of abrasive grains, their average rake angles, which can be as low as-60 °, their varying radial positions, and their extremely high slice pets, which are generally 30 m/s.

The high degree of dimensional delicacy and the face integrity of the produced work piece must be taken into account when erecting or installing the grinding process layout on the milling machine. It's important to take into account how machine tools can be modified to meet the requirements of the pottery grinding process. Disruptive rudiments like thermally convinced distortions and static stiffness of the grinding outfit should also be taken into account. To ensure steady and accurate machining processes, it's important to regulate both the grinding pressures and the grinding temperature. Grain force is commensurable to reuse variable if it's assumed that the cutting force on the grain is commensurable to the cross-sectional area of the under-formed chip.

Dehusking

Dehusking is the post-harvesting process of removing the husk and bran from the grains and producing polished grains that are mulled, free from contaminations, and contain a minimal number of broken grains (Pal *et al.*, 2020).

There are two approaches

a) **Traditional approach:** When pounding grain by hand, a rustic mortar with one or further deep pockets for storing it and one or two pastoral women folk use a Pestle, a 5- to 6- bottom-long rustic log. This system's strike is that it takes a lot of time and requires further mortal power to produce a product with a harmonious quality. Compared to rice milling machines, the grain is more nutritional. The system would affect a large quantum of cracked and fractured grains, and it would take several cycles to de-husk the grains and to some extent polish them. Rice can only be reused in little quantities at a time and it requires a lot of energy.

b) **Mechanical approach:** The hand-pounding system has drastically lowered since the arrival of newer mechanized and electrically operated manufactories since it could not contend with the volume and quality of reused rice produced by the machine manufactories. Wood is the primary element of mortar and pestle, and pastoral homes constantly utilised wood, which may contribute to deforestation and has an impact on our ecology.

Polishing process

The last steps in the manufacture of work components sometimes involve polishing processes. These procedures are now often performed manually, which poses a risk to the operator's health (Chaves *et al.,* 2015). When using soft pads and fine particle abrasives, polishing creates a surface that resembles a mirror (Marinescu *et al.,* 2000). The polishing abrasives are diamond powder with particles of less than 1μm.

A switch to a computer-controlled polishing approach is particularly advantageous since, as has frequently been noted; high-grade polishing requires a considerable degree of talent. Using a cylindrical shape mounted feeling tips and 0.25 m diamond paste, the hard-turned stainless steel pieces were automatically peripherally polished on a CNC machine. By using a computer controlled automatic and repeatable process, the goal of this polishing procedure is to achieve a super smooth cylindrical surface free of any scratches or turning marks, with a dimensional change in the range of 1-2 m on diameter.

Water jets

When they first arrived in the United States in the 1970s, water jets were only used for cleaning. New applications emerged as technology advanced to include abrasive water jets. But up until recently, the building sector has not made much use of this instrument. The water jet process is acknowledged as the most adaptable and fastest-growing process in the world (according to Frost & Sullivan and the Market Intelligence Research Corporation), in addition to being cost-effective. All across the world, high-production applications use water jets. They support various technologies including routers, milling, laser, EDM, and plasma. Water jet cutting doesn't involve any toxic gases or liquids, and the process doesn't produce any harmful vapors or compounds. A water jet-cut surface doesn't have any mechanical strains or heat-affected zones. It really is a productive, adaptable cold-cutting process (Dixit *et al* 2015) (Figure 4).

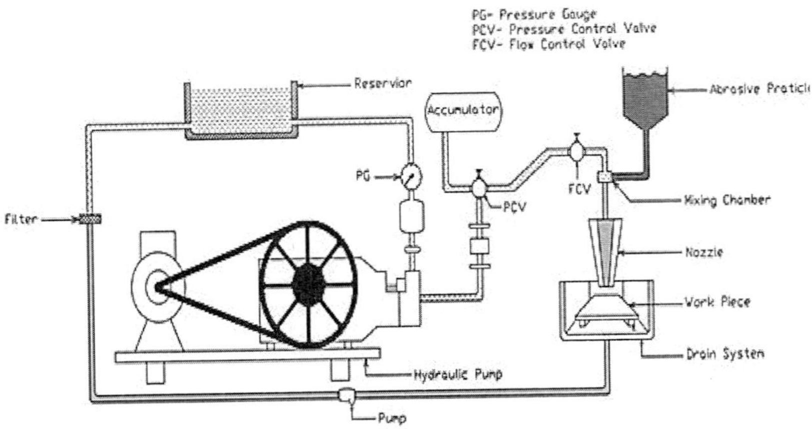

Fig 4: Water jets

Abrasion

The primary persecuted concrete blocks (periphery 5 40 mm) are ground with the help of a ball mill or rod mill so that it effectively removes the tenacious hardened cement pastes which can produce high-quality reclaimed summations. This fashion can be nominated as an abrasive grinding system. Treatments can be divided generally into two dry treatment or wet treatment. The working principle of the ball mill is shown in Figure 5. The ball mill contains a lot of sword balls between their partitions and these interactive collisions between sword balls and recycled summations or interactive collisions simply between recycled summations will enhance the junking of the tenacious cement paste, therefore perfecting the parcels of recycled summations. Still, some of the limitations associated with the abrasive grinding outfit include advanced energy consumption, lower effectiveness, and severe wear and tear and gash (Huang *et al.*, 2019).

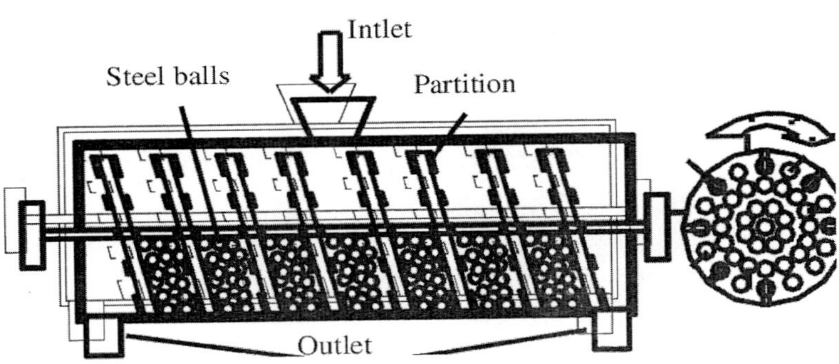

Fig 5: Ball milling equipment using abrasion grinding method

Friction

A general definition of friction is an outside force that prevents the motion of the other two forces. The friction that is produced between the tool and the knife can be used to understand how milling parameters are directly influenced. Therefore, the friction between the tool and the work piece could be used to understand milling settings. In addition to being dictated by the determination of the tangential element of the stress vector at the tool-to-chip interface, friction laws also play a role in the modeling of contact difficulties in machining. One of the simplest friction models is the Coulomb model. The law of disunion between interfaces is typically modelled by a modified Coulomb's law, with the disunion measure f assumed to be constant for a certain contact interface (Zarrouk *et al.,* 2021).

Flour milling machines

Roller mills and flour mill grinders are two examples of flour milling equipment. Other frequently used equipment includes a square plansifier, planetary mixer, and rotary sieve.

Pulse milling machines

Hammer mill

The size of the material supplied into the hammer mill is reduced by impact force. The feed hopper, feed screw, hammers, product exit, and screen discharge make up the hammer mill. It has a high-speed rotor within that rotates the hammer between 1500 and 4000 times per minute. The material is loaded into the feed hopper, where it is then sent to the hammer via the feed screw. The revolving hammers coupled to the fixed or swinging shaft provide kinetic energy, which is used to reduce the feed's particle size. Although the shear stress between the feed and screen and other pieces of the equipment contributes to the impact force. The screen size of the screen discharge is used to control the fineness of grinding. Through the product outlet, the screened particles can be collected.

Ball mill

A ball mill specific kind of grinder used for chopping, mixing, and grinding large materials. It primarily operates under the effect and attrition approach. Due to the impact as the balls fall close to the top of the shell, size reduction occurs during this process. Powders are mostly produced using this.

Impact mill

In impact milling, stress before rupture is concerned with a single point and this could be due to the impact created. This is typically used in cases of pulses.

Roller mill

The roller mills work by utilizing the compressive force produced by the rotating rollers. These mills have two sides, separated by separative drives, product feeding, and ground product pickup. Passage would be between 200 and 220 rpm. Each pair of the roll has a separate drive, where the front drive is driven by a belt.

Grinders

Grinders are generally power-operated machine tools, where substances are fed into constantly rotating abrasive wheel so that the thin layer is removed. Angle grinders, surface grinders and die grinders are some of the grinders used commonly. These are generally used to increase productivity.

Cutting machines

Cutting forces play an integral role in cutting machine. Machine performance is also related by the cutting properties such as hardness and toughness. End mill, face mill, ball cutter, slab mill, side-and-face cutter, involute gear cutter, fly cutter, hollow mill, shell mill, roughing end mill, dovetail cutter, and woodruff cutter are some of the common cutters employed for milling purposes (Meng *et al.*, 2017).

Oil expellers

Oil expeller can be defined as an instrument that undergoes mechanical extraction to produce oil from raw materials with the help of pressure by rotating the feed to the expeller screw manually or automatically. A mechanical method for the extraction of oil is the Screw press method. Low recovery could be one of the limitations of this technique. In screw press method consist of a hopper where seeds are fed and confined in the barrel and thus extraction is achieved. The pressure generated during this process is so high that it forces the oil from the seeds to come out. Finally, the container compresses the material, and extraction is achieved. And spent solids are removed as cakes. Continuous feeding of seeds helps the expeller screw to grind, crush and press the oil out as it passes through the machine. Heat in the range of 140-210°F (60-99°C) is created by pressure generated in expeller pressing (Khan *et al.*, 2016).

Machine efficiency

OEE (Overall Efficiency Effectiveness) could be described as one of the best parameters to evaluate machine efficiency. Effective and efficient upgrades made by the measurement of machine performance and production equipment could lead to OEE. Destruction of components, decreased production, lowered energy production, preparation length and adjustment time could be some of the factors affecting efficiency. Scheduled downtime, Update on real cycle times, reports of daily operator reports (daily schedule controls), and direct data retrieval in the field could sum up as some of the best indicators (Astari *et al.*, 2019).

Power requirement

A number of inserts in turn determines the power consumption. As the number of inserts increases power is also increased due to the feed velocity that is proportional to the number of inserts, where the feed rate per tooth and spindle speed are constant. More power is consumed and more material will be removed with higher feed velocity. Energy consumption can be reduced by low proportion of total power and by turning off the machine tool to a standby state. Thus, up-milling requires more energy than down-milling. Therefore, the milling modes of power consumption in descending order are up milling, down milling, and symmetric milling (Meng *et al.*, 2017).

Conclusion

Parboiled rice represents around 20% of all rice production output worldwide. The steeping, heating, and drying steps of the parboiling process significantly change the properties, milling, and cooking behavior of rice. The process of parboiling was developed to enhance rice quality. The rough rice is soaked, steamed, then dried during this process which increases milling yields, nutritional content, and resistance to insects and mold whereas the goal of milling process was to enhance product quality and nutritional value, encourage new product development, utilize the agricultural produce efficiently, lessen the cooking period and augment the palatability of food. Apart from the unit operations such as parboiling and milling, flour milling machines, pulse milling machines, grinders, cutting machines, oil expellers, machine efficiency, and power requirement were also discussed in this chapter.

References

1. Bhattacharya K.R. Improvements in technology for parboiling rice. *Cereal Foods World*, 2013; 58 (1), 23.
2. Buggenhout J, Brijs K, Celus I, and Delcour J.A. The breakage susceptibility of raw and parboiled rice: A review. *Journal of Food Engineering*, 2013; 117(3), 304-315.

3. Champagne, E. T., Bett-Garber, K. L., McClung, A. M., and Bergman, C. (2004). Sensory characteristics of diverse rice cultivars as influenced by genetic and environmental factors. *Cereal Chemistry*, 81(2), 237-243.

4. Dendy, D. A., and Dobraszczyk, B. J. (2001). *Cereals and cereal products: chemistry and technology*. Boom Koninklijke Uitgevers.

5. Derycke, V., Vandeputte, G. E., Vermeylen, R., De Man, W., Goderis, B., Koch, M. H. J., and Delcour, J. A. (2005). Starch gelatinization and amylose–lipid interactions during rice parboiling investigated by temperature resolved wide angle X-ray scattering and differential scanning calorimetry. *Journal of Cereal Science*, 42(3), 334-343.

6. Juliano, B. O., Perez, C. M., and Kaosa-ard, Mingsaran (1985). Grain quality characteristics of export rice in selected markets. *Consumer demand for rice grain quality*, 221-234.

7. Pillaiyar, P. (1988). Rice. Postproduction manual.

8. Campus, P. (2018). Year-I, Part-I. *Bachelor of Aerospace Engineering*, 24.

9. Chaves-Jacob J., Linares J.M., Sprauel J.M., (2015) Control of the contact force in a pre-polishing operation of free-form surfaces realized with a 5-axis CNC machine, *CIRP Annals - Manufacturing Technology* 64(1), 309–312.

10. Dixit, A., Dave, V., and Baid, M. R. (2015). Water jet machining: an advanced manufacturing process. *International Journal of Engineering Research and General Science*, 3(2), 288-292.

11. Floor Scrubber, archived from the or*iginal on* 2016-08-31, retr*ieved* 2008-01-04

12. Marinescu I. D., Tronshoff H. K., Inasaki I., (2000), Handbook of ceramic grinding and polishing.

13. Pal, S., Akhtar, J., Balki, S., and Parvez, M. S. (2020). Grain dehusking machines-a review.

14. Shneor, Y. (2018). Reconfigurable machine tool: CNC machine for milling, grinding and polishing. *Procedia Manufacturing*, *21*, 221-227.

15. Huang, K., Ding, T., Xiao, J., & Singh, A. (2019, August). Modification on Recycled Aggregates and its Influence on Recycled Concrete. In *IOP Conference Series: Earth and Environmental Science* (Vol. 323, No. 1, p. 012136). IOP Publishing.

16. Zarrouk T, Salhi J.E, Nouari M, Salhi M, Atlati S, Salhi N, & Makich H. Analysis of friction and cutting parameters when milling honeycomb composite structures. *Advances in Mechanical Engineering*, 2021; 13(7), 16878140211034841.

17. Chakraverty A, Mujumdar A.S, and Ramaswamy H.S. (Eds.). 2003. *Handbook of postharvest technology: cereals, fruits, vegetables, tea, and spices* (Vol. 93). CRC press.

18. Wood, J. A., and Malcolmson, L. J. (2021). Pulse milling technologies. In Pulse foods (pp. 213-263). Academic Press.

19. Khan M.H.U, Mondal D, and Hoque S. Design and construction of oil expeller press with structural analysis of screw with ansys. *PDm*, 2016. 50(4).

20. Astari, R., Surbakti, B., and Marizka, D. A. (2019, May). Overall Equipment Effectiveness Analyse for Performance of CNC Milling Machine Operation. In *IOP Conference Series: Materials Science and Engineering* (Vol. 505(1) p. 012139). IOP Publishing.

21. Meng, L., Zhang, C., Ren, Y., Luo, M., and Tian, G. (2017, August). Study on the power consumption of different milling modes and number of inserts in face milling processes. In *2017 13th IEEE Conference on Automation Science and Engineering (CASE)* (pp. 1475-1480). IEEE.

Chapter - 9

Bread - Ingredients - Additives - Chemical Leavening Agent - Different types of Bread - Methods of Bread Preparation - Bread Spoilage and Remedies - Quality Aspects of Bread and its Standards

Authors

Abinaya S

M.Sc. Scholar, Centre of Food Science and Technology, Chaudhary Charan Singh Haryana Agricultural University, Hisar, Haryana, India

Abirami K

M.Tech. Scholar, Department of Food Safety and Quality Assurance, College of Food and Dairy Technology, Koduveli, Tamil Nadu, India

Guruju Nookaratnam

M.Tech. Scholar, Department of Food Process Technology, Vasanthrao Naik Marathwada Krishi Vidyapeeth, Parbhani, Maharashtra, India

Chapter - 9

Bread - Ingredients - Additives - Chemical Leavening Agent - Different types of Bread -Methods of Bread Preparation - Bread Spoilage and Remedies - Quality Aspects of Bread and its Standards

Abinaya S, Abirami K and Guruju Nookaratnam

Abstract

The practice of baking bread dates back to quite primitive times throughout history. Breadmaking can prove to be an artistic endeavor, especially when working with yeast. Over thousands of years, artisan bakers have come up with conventional bread with different varieties by applying their skills by utilizing flour from wheat grain as a base ingredient. Now, the increasing consumer demand for healthier choices expands the development of specialty bread such as multigrain bread, high-fiber bread, and low-calorie bread, etc. This chapter provides concise information about the role of essential ingredients in bread making and their functions. An overview of the primary bread-making methods is discussed, along with how they impact the quality of bread. The primary objective of any leavener is to release the gas that causes bread to expand during fermentation. The widely known biological leavening (yeast) and various other chemical leavening agents and their functions were discussed. The term “Quality” refers to the degree of excellence, hence the quality characteristics of bread and associated changes during storage and their control measures are considered.

Keywords: Quality, wheat, dough development, bread staling, leaveners.

Introduction

The term "baked products" refers to a variety of goods, including bread, crackers, cakes, cookies, and pastries that involve heat processing and utilize wheat flour as a key ingredient. In 2023, the bread industry will generate US$0.45 trillion in revenue. The market is anticipated to expand at 5.87% yearly (CAGR 2023-2027) (Statista, 2023). Bakery products, especially bread have played a significant role in diets and have remained a staple food throughout the history of civilization. Till date, the bottom shelf of the food pyramid is comprised of bread and other cereal products, and intake of these products is advised in all dietary recommendations (Rosell, 2011). Bread plays a vital role in nutrition. It is a major repository of protein, complex carbohydrates (mostly starch), dietary fiber, vitamins (particularly B vitamins), minerals, and energy. The two most common cereals suitable for breadmaking are wheat and rye. However, wheat is a special cereal that could be potentially used to make a variety of leavened bread that satisfies consumer preferences around the world (Rosell, 2011). The unique characteristics of bread are mostly influenced by the added ingredients. The primary ingredients for the bread making were flour, water, salt, and yeast. Other ingredients such as sugar, fat, dough improvers and emulsifiers were added to attain the desired quality attributes of the bread. An important aspect of bread is the development of a gluten network and the inclusion of air bubbles within the dough structure (Cauvain, 2012). This gluten network plays a vital role in trapping gas produced during fermentation and assists in the development of crumb structure which imparts enjoyable eating quality as well as texture. There are numerous methods for the preparation of bread, all of which attempt to turn wheat flour and other components into food that is light, fluffy, and delightful. The overall quality of the bread is decided by the complex associations between the ingredients used, their quality and quantity, and the processing methods used to produce dough.

Ingredients

The basic ingredients for any bread formulation are flour, water, leavening agents (especially yeast), and salt. Other ingredients such as fat, sugar, milk, eggs, etc are also added to obtain the desired color, flavor, nutrition, and crumb structure. Each of the basic ingredients and its functions were listed below;

1. Flour

Cereals contain starch and proteins which provide structure to the bread. The yield of the bread depends upon the type and quality of the flour. Most often wheat flour of hard variety is used for bread making because it contains 68-76% starch and 10 to 12% protein. Wheat protein, (gluten), which is made up of gliadin and glutenin, is responsible for the viscoelastic properties of the dough when flour is mixed with water. Gliadin is responsible for loaf volume and viscous nature of gluten, whereas, glutenin provides strength and elasticity to dough. The gluten network formed during dough preparation retains air produced during fermentation thereby influencing the yield and quality of the bread. Therefore, wheat flour with high protein gives a higher yield and vice-versa. Starch and sugars present in flour act as a nutrient source for yeast, thereby reducing the need for the addition of yeast food to the dough.

2. Water

Water is an important part of bread, which is necessary for the development of gluten and helps the starch to swell. It blends with the flour to produce sticky and elastic dough. It dissolves and activates the yeast. The amount of water needed for the dough depends on the protein content of the flour. In some enriched bread formulas, such as brioche, milk is used instead of water.

3. Salt

Salt plays a critical role in bread making by influencing the rheological properties. It strengthens the gluten network so that it can hold the air strongly and helps in rising bread. Salt limits the action of the yeast and thus regulates the fermentation. Moreover, it also improves the flavor of the bread. Insufficient/ absence of salt causes the dough to become weak and sticky as a result of poor gluten network formation caused by increased water-protein interactions and decreased protein-protein interactions (Avramenko *et.al,* 2018).

4. Leavening agents

These are the components that increase the volume of the dough by producing carbon dioxide gas through the action of acid in the presence of water. The gas bubbles produced get entrapped in the gluten network of the dough, rise along with it, and thus increase the volume of the bread. The leavening action is responsible for good volume, improved eating quality, and a uniform cell structure. There are three types of leavening agents includes biological (yeast and sourdough starter), chemical (Baking soda, baking powder, ammonium bicarbonate, and potassium bicarbonate), and mechanical (creaming, whisking, and kneading, etc) leavening agents. The most common leavener used from age-old practice is yeast. Each of these leaveners has its specific characteristics and confines which render it appropriate for various products (Hui *et al.,* 2008).

Additives

A food additive is any substance intentionally added to food for technological purposes such as to improve shelf life, and maintaining its nutritional qualities and sensorial attributes (taste, texture, appearance). The following additives are commonly used in bread making;

- **Oxidizing agents:** Oxidants play a crucial part in bread-making by enhancing the dough's capacity to retain gas via improved gluten formation and also shortening the dough development time. The commonly used oxidizing agents in bread and related products are potassium bromate (slow-acting), potassium iodate, calcium iodate, and azodicarbonamide (fasting-acting) (Gioia *et al.*, 2017).
- **Reducing agents:** It weakens the strength of the dough by disintegrating the cross-linkages within the amino acid chains that make up the gluten network and improves the elastic property of the dough. The most commonly used reducing agent is L-cysteine (an amino acid) (Chan, 2015), because it improves the machinability, speed up the mixing process, also shortened the proofing time. Other examples include inactivated yeast and enzymes like proteases.
- **Emulsifiers:** These are the substances that stabilize the mixture of two immiscible compounds. A variety of emulsifiers may be used to enhance the quality of bread, each of which has a unique action and specific functions. The widely used emulsifiers in the bread industry are categorized into two groups based on their function i) crumb softener (eg: Monoglycerides and diglycerides), and ii) Dough conditioner (eg: Diacetylated Tartaric Acid Esters of Mono- and Diglycerides of Fatty Acids (DATA), and polysorbates) (Gioia *et al.*, 2017). Other examples include sodium stearoyl lactylate (SSL), Glyceryl monostearate (GMS), and lecithin. The main function of emulsifiers in bread making are i) stabilization of dough, ii) speed up the rate of hydration, iii) acceleration of the inclusion of tiny air bubbles during the kneading process iv) Better volume of loaf and crumb, and v) prevent staling of bread (Moonen & Bas, 2015).
- **Enzymes:** These are generally proteins in nature and serve as biological catalysts. In bread making, enzymes are used as fermentation enhancers, dough conditioners, anti-staling agents, and crumb whiteners. The two well-known classes of enzymes widely used in bread making are alpha-amylases and proteinases (Stanley *et al.*, 2007). Other enzymes such as hemicellulase, lipase, glucose oxidase, and lipoxygenase are also commonly used.
- **Preservatives:** Weak organic acids like propionic, benzoic, sorbic, and their salts are commonly employed in bread making as chemical preservatives to inhibit the growth of microbes and extend the keeping quality (Chan, 2015). Propionates are very effective against molds at a concentration of 0.1- 0.2%, especially calcium propionate.

Further, sorbic acid at a concentration of 0.125% to 0.3% is effective at preventing mold spoilage in baked goods but it is limited in bread and yeast-leavened products, because of the adverse impact on the qualities of the dough rheology and the activity of yeast (Stanley *et al.,* 2007). Apart from chemical preservatives, some of the natural/ bio-preservatives are of commercial interest. The best-known example is Lactic acid bacteria, which is effective against mold.

Chemical leavening agents

Chemical leavening agents are combinations of chemical substances that react with one another, and in the presence of moisture or heat, liberate gases. Most of the chemical leavening agents are combinations of low molecular weight inorganic acids and salts of bicarbonate. The most commonly used chemical leavening agents include baking soda (sodium bicarbonate), baking powder (a combination of bicarbonates and acids), ammonium bicarbonate, and potassium bicarbonate were discussed as follows;

- **Baking soda:** It is also known as sodium bicarbonate ($NaHCO_3$), decomposed into sodium carbonate, water, and carbon dioxide when heated. It acts readily with heat if used singly and liberated carbon dioxide escaped quickly from unleavened dough. A few drawbacks exist with using sodium bicarbonate, it only dissolves to a maximum of about 25%, and the sodium carbonate and residue left combine to produce an alkaline environment that produces a yellowish tinge on the surface and the crumb of the bread (Neeharika *et al.,* 2020). Hence, baking soda must be used with an appropriate amount of acid to produce a neutral residue.

- **Baking powder:** It is a combination of sodium bicarbonate, an acid ingredient, and filler (usually corn starch). The acid ingredients generally used are cream of tartar, lactic acid, monocalcium phosphate, sodium aluminium silicate, and sodium acid pyrophosphate (Brodie & Godber, 2007) and the purpose of fillers is to standardize the amount of baking powder used and also prevent premature reaction between the soda and acid. Industrially used baking powders produce a minimum of 12.0% available CO_2 gas by weight, whereas, household powders produce 14.0% of CO_2 gas (Neeharika *et al.,* 2020). Like baking soda, the sodium bicarbonate and acid salts present in baking powder are combined with water and liberate CO_2 gas. Baking powders are generally classified based on the use of acid ingredients and the rate of the reaction. According to the rate of the reaction, they are categorized into i) single-acting (Baking powder with only one acid ingredient) ii) slow acting (requires high temperature (baking) to liberate CO_2 gas) iii) Fast acting (liberate much CO_2 gas during tabletop preparation and little during baking), and iv) Double acting (Combined action of slow and Fast acting baking powders). The composition and proportions of sodium bicarbonate and acid ingredient must be adjusted to ensure that i) a predetermined quantity of carbon dioxide is released, ii) the controlled rate of reaction, and iii) no residues are left upon completion of the reaction (De Leyn, 2014).

- **Ammonium bicarbonate (NH_4HCO_3):** Generally, ammonium bicarbonate decomposes at a temperature over 60°C and liberates ammonia, CO_2, and water. Both ammonia and CO_2 evaporate as gases, leaving the baked goods residue-free. Since it has no impact on the pH of baked items and does not need acid salts. It is most commonly used in less moist foods. If the moisture content of the product is above 5 %, it produces an undesirable ammonic flavor and renders the product unpalatable, because of the water-soluble nature of the ammonia gas liberated.
- **Potassium bicarbonate ($KHCO_3$):** It is generally used when there is a reduction in the level of sodium content in baked products. It also decomposes in the presence of heat to liberate CO_2 gas, but for the thorough decomposition, it needs an acid ingredient as well as a prolonged baking process. Because of its slow-acting nature, it produces products with a smooth and fine crumb structure. A few drawbacks exist with using potassium bicarbonate, incomplete decomposition results in a product with a bitter taste; also highly hygroscopic nature of $KHCO_3$ renders storage problematic (Neeharika *et al.*, 2020).

All these chemical leavening agents must be stored in air-tight containers and kept under cool and dry place, if it is not properly sealed, due to their hygroscopic nature; it absorbs moisture from the surroundings and gives up their leavening capacity.

Methods of bread preparation

The process of bread making involves a sequence of different steps as follows; i) Scaling ingredients ii) Mixing/ Preparation of dough iii) Fermentation iv) Folding v) Dividing vi) Rounding vii) Intermediate proofing viii) Panning ix) Proofing x) Baking xi) cooling and xii) storage (Gisslen,

2013). All of the steps mentioned above, except making the dough, are the same for all types of bread; however, the variations in the preparation of dough produce uniqueness in different types of bread. There are various methods of bread making (dough making), which includes;

1) **Traditional bulk fermentation method:** The important aspect of the traditional dough development method is the prolonged fermentation time. The traditional method is categorized into three types based on the method of mixing and the ingredients. They are as follows;

   a) **Straight dough method:** This is the fundamental traditional method of preparing bread, from which all other methods are developed. This method involves a single mixing step following the addition of the required amount of water and other ingredients to the flour. The mixing time varies between 15-30 minutes, but it may vary based on the capacity of the mixer and production size. Mixing needs to proceed until the dough becomes elastic with a smooth, silky, and non-sticky finish. The dough mixture is then held for bulk fermentation, often for 1-2 hours. It is simple; less time-consuming compared to the sponge and dough method, rather, the initial investment is high.

   b) **Sponge and dough method:** This method consists of two stages: i) Add a portion of the liquids, yeast, and flour to make a soft dough and ferment until the volume doubles, ii) Fold and then combine the remaining ingredients, mix well to form a soft and elastic dough. It provides bread with a higher volume and desired crumb structure, flexible in terms of mixing and fermentation, whereas the losses during fermentation are high (Gisslen, 2013).

   c) **Delayed addition of salt**: In this method, the whole quantity of flour required for a single batch is added during the initial mixing phase, and fermentation proceeds for 3 hours. Following fermentation, the dough is subjected to a second mixing, during which all the salt is added at the final stage of the mixing process. The purpose is to improve the structure of gluten and allow complete hydration of the flour protein (gluten) prior to the addition of water-binding salts (Stanley *et al.*, 2007).

2) **Chemical dough development method:** In this method, after completion of the mixing process, the dough receives very brief (less

than 1 hour) or no fermentation. Here, the dough development process is carried out through chemicals, especially reducing agents (E.g. L-cysteine hydrochloride, sodium metabisulphite). This method requires less energy and time relative to traditional methods, and no need for expensive specialized mixing equipment. The major drawback is that bread developed by this method lacks flavor and aroma compared to the traditional method.

3) **Mechanical dough development method:** In this method, the desired changes obtained by the traditional method are carried out in the mixers by the use of intensive mixing and high energy. The most widely used one is the Chorleywood Bread Process (CBP), which was developed in the United Kingdom. The primary distinction between the CBP and traditional method is the rapid rate with which the dough develops in the mixer as opposed to an extended resting period. The preparation of dough involves the addition of flour, liquids, and other ingredients in the mixer and the rate of mixing is about less than 5 minutes (Mondal & Datta, 2008).

Types of bread

The production of specific varieties of bread has been pursued in different regions around the world. Such varieties are developed by modifying the following aspects (a) incorporating different/ functional ingredients (b) modifying ingredient formulation (c) modifying processing methods (d) modifying shape and size, etc. Additionally, the uniqueness of the specific varieties of bread obtained from the gluten (wheat protein) formation and the artistic ability of the bakers. There are three different types of bread, which were discussed below;

i) **Leavened bread**

- **Whole wheat bread:** It is widely acknowledged that whole-wheat flour has considerably better nutritional value than the more frequently used refined white flour. Hence the bread baked from whole wheat flour is more nutrient dense with an instinctive and pleasing aroma of wheat.
- **White bread:** The bulk of bread and other baked goods are generally produced using refined white wheat flour, which is devoid of the outer bran and germ layers which contains a significant amount of vitamins, minerals, and phytonutrients (Collar, 2016).
- **Brown bread:** Typically, brown bread is produced with an equal

proportion of white flour and whole wheat flour. Additionally, to attain desired flavor and brown color, brown sugar and caramel were added.

- **Rye bread:** Whole, steel-cut, crushed, or milled rye kernels that may also have been malted are used for the preparation of rye bread. Rye, a cereal grain is an ideal source of dietary fiber along with some important phytonutrients such as lignans, phenolic acids, and alkylresocinols, which are proven to reduce several chronic diseases (Kamal-Eldin *et al.,* 2008).
- **Hearth bread:** Refers to a traditional baking practice in which a portion of fermented dough is baked on the bottom of a wood-fired oven.
- **Baguette:** It is also known as French or Vienna bread. It is whiter, lightweight, and sweet relative to the ancient sourdough bread method.
- **Pan bread:** Originated in England and Poland, refers to bread made in a baking pan.

ii) Unleavened/ Flatbread

These are easy-to-make breads using flattened dough, they are prepared from several kinds of grains, particularly wheat, maize, sorghum, finger millet, and teff (Collar, 2016). Flatbreads are generally categorized into three types i) Single-layered (E.g.: chapatti, parotha, tortillas, and Arepa), ii) Dough-layered (E.g.: Barbari, Nan, Tandoor, and Pita), and iii) Batter-layered (E.g.: Dosai, injera, Crepe, pancake). These products are usually characterized by their acidic (sour) flavor and leavened texture due to the mixed fermentation of lactic acid bacteria and yeast.

iii) Steamed bread

Steamed bread/ buns are popular all over south East Asia, especially in China, Japan, and Korea. The most common varieties of steamed bread and bun are Bao (filled) and, Mantou (unfilled). The only well-known steamed bread is Mantou, which is unfilled steamed bread, whereas Bao comes under steamed buns and is generally filled with pork belly, which is characterized by its white exterior that breaks out during steaming, considered to be a unique characteristic of the Bao (Collar, 2016).

Bread spoilage and remedies

It is evident that particular quality attributes are necessary for specific bread variations to be liked by the maximum number of consumers (Cauvain,

2012). For instance, pies must have a flaky, soft crust that is golden brown around the edges; otherwise, we would reject the product. The quality and quantity of the ingredients used in the formulation, as well as the methods used to prepare the dough, all have a major influence on the quality of bread. Bread is distinguished by having a dry, thin crust that is surrounded by the outer surface and a relative firmness that yet maintains springiness and chewiness for the breadcrumb structure (Cauvain, 2012). These quality attributes deteriorated gradually after baking and during storage, accompanied by a range of modifications, including, physical (staling, redistribution of moisture, texture and flavor loss) (Smith *et al.,* 2004), chemical (development of rancidity and off-flavor) and microbial (bacterial, fungal and yeast) spoilage (Rahman *et al.,* 2022).

Physical spoilage

Staling

The term "staling" indicates the gradual decline in the acceptance of the bread by the consumers as a result of physical and chemical modifications that take place in the crust and crumb while storage, which excludes microbial spoilage. The staling of bread is primarily correlated with the firming of the crumb, and explained by two distinct processes; a) the transfer of moisture from crumb to crust influences the firmness of the crumb and, b) the inherent firming of the cell wall is related to the re-crystallization of starch during storage of bread (Stanley *et al.,* 2007). The staling of bread is categorized into two types; i) Crust staling and ii) Crumb staling.

i) Bread crust staling

During the storage of bread, there is a moisture migration from the crumb to the crust, which results in increased moisture content of the crust. The bread crust becomes soft and fibrous (Gray & Bemiller, 2003). In general, the crust has an original moisture content of about 12% and is easily moistened by the interior crumb, which has a moisture content of roughly 45% (Stanley *et al.,* 2007). However, compared to crumb staling, crust staling is less unacceptable.

ii) Bread crumb staling

The main constituent of flour, or roughly 75% of the total weight, is made up of starch, which is essential for the structure, physical characteristics, and quality of bread. The staling of bread crumbs was highly correlated with the gelatinization and retrogradation nature of starch granules. In general, the granules of native starch are not soluble in cold water, yet they readily absorb water and swell when heated in an aqueous solution. As the temperature of the aqueous medium increases, the starch granules begin to vibrate more violently, thereby disrupting intermolecular associations and permitting their hydrogen bonding regions to interact with more water molecules, thereby disintegrating starch granules. As a result, there is a reduction in the size and the number of crystalline areas. The crystalline structure of starch granules generally softens at a temperature range between 60 and 90 °C in baked goods. The condition of starch granules in the final baked bread might be completely gelatinized to total loss of crystalline structure with the release of some polymers. The liberated polymers as well as the amylopectin within the granules impart a rubbery texture to the bread crumbs (Stanley *et al.,* 2007). Further, the most common cause of staling is the time-dependent re-crystallization of amylopectin, which transforms it from an entirely amorphous state when first cooked to a partly crystalline state when it is stale. The modifications in the physical alignment of the branched chain amylopectin molecules within the swollen starch granules lead to increased hardness of the bread crumbs.

The staling of bread crumbs is not only associated with the starch granules but also due to the protein fraction (gluten) of the flour. The interactions of hydrogen bonds between the hydroxyl group of the starch granules and the amino group of the protein fractions gradually increased and attained strength during storage which results in the loss of kinetic energy of the bread crumbs and a further increase in the firmness of the crumb (Martin *et al.,* 1991).

Control measures: The use of anti-staling substances including enzymes (especially alpha-amylases), emulsifiers (momglycerides), alcohols (produced during fermentation of bread), pentosans (non-starchy polymers), and sugars (sucrose, fructose, and glucose) might inhibit or slow down the process of staling. Additionally, deep freezing has also been found to hinder microbial activity and the staling process (Stanley *et al.,* 2007).

Flavor loss

The flavor of bread comprises a variety of senses including aroma, taste, warmth, sight, and eating qualities. During storage, the flavor constituents of the bread might dissipate or supersede. The typical strong dough and bran aroma and flavor of fresh bread loaves have given way to a dusty, aged, and rancid aroma and flavor and sharp taste, together with a strong astringent and puckering feeling throughout storage (Fadda *et al.,* 2014).

Chemical spoilage

The most common chemical spoilage of baked foods, especially those rich in fat is "**Rancidity**", is described by the process of breakdown of fats and oils in the presence of oxygen, light, catalysts, moisture, or microbial action which results in the production of short-chain molecules such as free fatty acids, aldehydes, and ketones. The emergence of an off flavor and odor are typical indicators of rancidity. The unsaturated fatty acids which contain more double bonds are highly susceptible to rancidity than saturated fatty acids.

Control measures: The use of antioxidants such as Butylated hydroxy anisole (BHA) (0.02%), Butylated hydroxytoluene (BHT) (0.02%), Tertiary Butyl Hydroxy quinone (TBHQ) (0.02%), propyl gallate (0.01%), citric, gallic and tartaric acid (0.01%), ascorbyl palmitate (0.02%) were effective in preventing auto-oxidation of lipids (Srivastava & Haridas Rao, 1997).

Microbial spoilage

The common spoilage of bread is usually "**Moldiness**" and "**Ropiness**". Moldiness is generally due to the growth of filamentous fungi or molds, whereas ropiness is due to bacterial growth, especially species of bacillus (Stanley *et al.,* 2007). More spoilage of bread is caused by mold than bacteria.

Moldiness

The common genera of fungi associated with the spoilage of bread include *Aspergillus, Penicillium, Cladosporium, Endomyces, Rhizopus, Mucor, Monilia,* and *Fusarium* (Dal Bello *et al.,* 2007). The other possible source of spoilage includes the use of contaminated raw ingredients, bread contact surfaces (equipment, utensils), and packaging materials (Reale *et al.,* 2013) also pave the way for the entry of pathogenic microbes that are detrimental to the quality of the baked goods.

- ***Pencillium*:** The development of greenish-blue color on the surface of the bread is identified by the growth of *Pencillium* species. The species of *Penicillium* that might cause bread spoilage includes; *Penicillium roqueforti, Penicillium paneum, Penicillium brevicompactum*, and *Penicillium chrysogenum* (Garcia *et al.,* 2019).
- ***Aspergillus*:** The species of *Aspergillus* generally produce greenish or purplish brown to black colonies on the surface of bread. The common species of *Aspergillus* that might cause bread spoilage includes; *Aspergillus niger* (Black), *Aspergillus flavus* (Olive green), *Aspergillus glaucus* (Pale green), *Aspergillus candidus* (Creamy) (Seiler, 1992).
- ***Rhizopus*:** *Rhizopus stolonifer*, which is also called bread mold, is the most common species of *Rhizopus*, produces white cottony

mycelium and black sporangia rendering it an extremely fluffy appearance on the bread surface is an indication of the growth of this species.

- ***Monilia***: *Monilia sitophila* (formerly *Neurospora sitophila*) is favored by high humidity and warm conditions, and produces pinkish colonies on the surface of the bread (Stanley *et al.*, 2007).
- ***Endomyces***: *Endomyces fibuligera* and *Trichosporon variable* (Yeast-like fungi) causes "chalky bread" by the formation of white, chalk-like colonies on the bread, but is rare (Frazier & Westhoff, 2016).

Ropiness

Ropiness is generally caused by bacterial species favoured by high humidity (more than 90%). The formation of ropy bread is mainly due to the growth of mucoid variants of *Bacillus* species such as *Bacillus subtilis* or *Bacillus licheniformis*. The spores of these species generally survive under baking temperature. The cause of the ropy state in the baked bread was thought to be the capsulation of bacteria along with the hydrolysis of starch and protein (gluten) fractions of the bread flour by the enzyme amylases and proteinases, respectively. The ropy region becomes soft and sticky and the color turns from yellow to brown (Frazier & Westhoff, 2016).

Control measures

The use of preservatives such as organic acids especially, propionic acids and their salts at a concentration of 0.1-0.2 % were effective against mold and bacteria that cause ropy bread. In addition to propionates, the use of ethyl alcohol as a spray on the surface of the bread loaves at a concentration of 0.5-3% extends the keeping quality of the bread and also acts as an antimicrobial agent (Legan, 1993). Other preservation methods such as the use of modified atmospheric packaging, and ionizing radiations like UV rays, infrared, and microwaves were found to inhibit the spoilage caused by microorganisms.

Quality aspects of bread and its standards

Bread quality is a highly individualized concept that significantly relies on the perception of consumers. The general understanding of bread quality could be consolidated by instrumental approaches, those that can be determined objectively, sensory attributes, including descriptive qualities that concern consumer perceptions, and nutrient characteristics associated with the efficiency and healthfulness of the bread products (Rosell, 2011).

General requirements: The crust must be devoid of blisters, pleasant golden, light-brown color, must not be charred, and devoid of ash and other extraneous particles. The perfect crust must be thin and split readily, and must not be fibrous, stiff, or thick. The volume of the bread is determined based on the volume-to-weight ratio, if the ratio falls within 5.75 to 1, it is considered as bread with good volume. The breadcrumb must have thin cell walls, small, well-spaced pores, and a bouncy texture. The flavor of the bread must be reminiscent of freshly baked bread and devoid of any unpleasant flavors or odour, such as staleness or sharpness. There must be no signs of ropiness or moldiness on the bread. The internal structure of bread must have thin walls which are distributed uniformly all over and must have a soft and silky texture without any signs of flakiness (ICS 67.060, 2020).

Microbiological limits: The presence of total viable count and yeast and mold should be 10^4 and 10^3 cfu/g (colony forming units per gram), whereas *Salmonella* and *Escherichia coli* must be absent (ICS 67.060, 2020).

Table 1: Standard requirements for different types of bread:

Products	**Moisture content (%) (max)**	**pH of aqueous extract**	**Acid insoluble ash, % (dry basis)**	**Crude fiber, % (dry basis)**	**Fat,% (min)**	**Total solids, % (min)**
Whole wheat bread	40	5.3 – 6.0	0.2	1.0	0.7	60
Brown bread	40	5.3 – 6.0	0.2	0.6	2	60
White bread	40	5.3 – 6.0	0.2	0.3	0.7	60

ICS 67.060, 2020; min – minimum; max- maximum; % - percentage.

Conclusion

Bread has long been considered as a staple food and consumed throughout the world as it is a significant source of energy, protein, and other micronutrients. Bread baking involves combining flour and other ingredients with water to form a dough, leavened with yeast that releases carbon dioxide, and baked to strengthen the solid protein foam produced, giving an extensible porous structure. Growing demand for nutritional and healthier choices can be met by bread because it is a good carrier of micronutrients as well as bioactive compounds. The quality of bread is influenced by various factors, and there exist tremendous possibilities for advancement. Such advancements should be based on intricate relationships that exist between components, composition, and processing methods. Meanwhile, baking equipment will keep on improving without affecting the quality of the final product. Bread baking is an art, skill, science, and technology that has been continued from ancient times and uses different ingredients and processing methods to develop specific types of bread through extensive research.

References

1. Avramenko, N. A., Tyler, R. T., Scanlon, M. G., Hucl, P., & Nickerson, M. T. (2018). The chemistry of bread making: The role of salt to ensure optimal functionality of its constituents. *Food Reviews International, 34*(3), 204-225.

2. Brodie, J., & Godber, J. (2007). Bakery Processes, Chemical Leavening Agents. *Kirk-Othmer Encyclopedia of Chemical Technology*, John Wiley & Sons, 1-14

3. Cauvain S. Breadmaking: an overview. *Breadmaking*, 2nd Edition, Woodhead Publishing, 2012; 9-31.

4. Chan, P. N. A. (2015). Chemical properties and applications of food additives: Preservatives, dietary ingredients, and processing aids. *Handbook of Food Chemistry*, 75-100.

5. Collar, C. (2016). Bread: Types of bread. *Encyclopedia of Food and Health,* 500-507.

6. Dal Bello F, Clarke C.I, Ryan L.A.M, Ulmer H, Schober T.J, Ström K, & Arendt, E. K. (2007). Improvement of the quality and shelf life of wheat bread by fermentation with the antifungal strain Lactobacillus plantarum FST 1.7. *Journal of Cereal Science*, *45*(3), 309-318.

7. De Leyn, I. (2014). Other leavening agents. *Bakery products science and Technology*, 175-181.

8. Fadda, C., Sanguinetti, A. M., Del Caro, A., Collar, C., & Piga, A. (2014). Bread staling: Updating the view. *Comprehensive Reviews in Food Science and Food Safety*, *13*(4), 473–492.

9. Frazier W.C, & Westhoff D.C. Food microbiology. *Contamination, preservation, and spoilage of cereals and cereal products*, 5th Edition, Mc Graw Hill Education, Chennai, 2016; 163-174.

10. Garcia M.V, Bernardi A.O, & Copetti, M.V. (2019). The fungal problem in bread production: Insights of causes, consequences, and control methods. *Current Opinion in Food Science*, *29*, 1-6.

11. Gioia, L. C., Ganancio, J. R., & Steel, C. J. (2017). Food additives and processing aids used in breadmaking. *Food additives*, 147-166.

12. Gisslen, W. (2013). *All Access Pack for Professional Baking, 6th Edition,* John Wiley & Sons, 107.

13. Gray J.A, & Bemiller J.N. Bread staling: Molecular basis and control. *Comprehensive Reviews in Food science and Food safety*, 2003. *2*(1), 1-21.

14. https://www.statista.com/outlook/cmo/food/bread-cereal-products/bread/worldwide

15. Hui Y.H, Corke H, De Leyn I, Nip W.K, & Cross N.A. *Bakery products: science and technology*, 1st Edition, John Wiley & Sons, 2008; 245- 265.

16. International Classification for Standards (ICS 67.060). (2020), 3rd Edition, Bread – Specification, Malawi Bureau of Standards.

17. Kamal-Eldin, A., Aman, P., Zhang, J. X., Knudsen, K. E. B., & Poutanen, K. (2008). Rye bread and other rye products. *Technology of Functional Cereal Products*, 233-260.

18. Legan, J.D. (1993). Mould spoilage of bread: The problem and some solutions. *International Biodeterioration and Biodegradation*, *32*, 33.

19. Martin, M.L., Zeleznak, K.J., & Hoseney, R.C. (1991). A mechanism of bread firming. I. Role of starch firming. *Cereal Chemistry*, *68*, 498–503.

20. Mondal, A., & Datta, A. K. (2008). Bread baking– A review. *Journal of Food Engineering*, *86*(4), 465-474.

21. Moonen, H., & Bas, H. (2015). Mono-and diglycerides. *Emulsifiers in food technology*, 2nd Edition, Wiley Blackwell, 73-92.

22. Neeharika, B., Suneetha, W. J., Kumari, B. A., & Tejashree, M. (2020). Leavening agents for food industry. *International Journal of Current Microbiology and Applied Sciences*, *9*(9), 1812-1817.

23. Rahman, M., Islam, R., Hasan, S., Zzaman, W., Rana, M. R., Ahmed, S., & Sunny, A. R. (2022). A Comprehensive review on bio-preservation of bread: An approach to adopt wholesome strategies. *Foods*, *11*(3), 319.

24. Reale A, Di Renzo T, Succi M, Tremonte P, Coppola R, & Sorrentino E. 2013. Microbiological and fermentative properties of baker's yeast starter used in breadmaking. *Journal of Food science*, *78*(8), 1224-1231.

25. Rosell, C. M. (2011). The science of doughs and bread quality. In *Flour and breads and their fortification in health and disease prevention*, Academic Press, 3-14.

26. Seiler, D. (1992). Basic guide to product spoilage and hygiene. Part 1: *Mould growth on bread*. Chorleywood Digest No. 121, CCFRA, Chipping Campden, UK, 117.

27. Smith J.P, Daifas D.P, El-Khoury W, Koukoutsis J, & El-Khoury A. Shelf life and safety concerns of bakery products – A review. *Critical Reviews in Food science and Nutrition*, 2004; *44*(1), 19-55.

28. Srivastava A.K, & Haridas Rao, P. (1997). Importance of additives in improving the quality of bakery products. *Indian Miller*, *28*(3), 15-19.

29. Stanley P, Cauvain S, & Young L. (2007). Technology of bread making, Springer New York, 240-261.

Chapter - 10

Development of New Cereal-Based Products and Development of Cereal Product Quality

<u>**Authors**</u>

Riddhi Rane

Student, BTech Food Science and Technology, School of Biotechnology and Bioinformatics, D. Y. Patil Deemed to be University, Navi Mumbai Maharashtra, India

Rohit Palghadmal

Ph.D. Scholar, Mahatma Phule Krishi Vidyapeeth, Rahuri, Maharashtra, India

Chapter - 10

Development of New Cereal-Based Products and Development of Cereal Product Quality

Riddhi Rane and Rohit Palghadmal

Abstract

Except for wheat, cereals have not been fully exploited in the food industry. As consumer preferences have switched to healthy foods, many new cereal-based products are now being formulated and the traditional cereal-based products are also gaining recognition. The developed products such as snacks (snack bar, extruded snacks, fish-cereal based snacks), beverages (Boza, Ogu), fermented products (Yosa) have high nutritional value. They have the potential to become a part of our healthy lifestyle. However, some of the products developed have a poor sensory profile (e.g., texture, taste, volume) which affects the consumer acceptance. This led to the need of developing new technologies and methods for improving the quality of these products. The quality can be developed either with the help of enzymes and yeast or by technologies such as DIC and dry-crystallization.

Keywords: Cereals, snacks, quality, DIC, dry-crystallization

Introduction

Milled wheat and oats were among the earliest cereal products available for human consumption. These had to be cooked before consumption. Advances such as precooking techniques or additives have resulted in reduced in-home processing requirements. Additionally, emergence of microwave oven has made eating these cereals more convenient (Tribelhorn, Handbook of Cereal Science and Technology, 1991).

However currently, owing to the increased desire for innovative and healthful foods along with rising number of diseases such as celiac in the population, attention has been shifted to gluten-free cereal-based products (Arendt and Dal Bello 2008; Salovaara *et al.* 2011). Apparently, oat, rice, corn, sorghum, millet, quinoa, amaranth and buckwheat have become popular to be used as basic ingredients or in combinations for preparation of gluten-free products with varying nutritional values and sensory acceptability

(Flander *et al.* 2007; Rosell and Marco 2008; Taylor *et al.* 2006; Angioloni and Collar 2011, 2012; Alvarez-Jubete *et al.* 2010; Rosell *et al.* 2009; Guiné & Correia, 2013).

This can be seen in the biscuit sector where the traditional recipes have been switched with new recipes which include cereals, pseudocereals, or other constituents like seeds to comply with the consumer needs for low-sugar, gluten-free, high nutrient products. Nevertheless, other factors such as the sensory profile and formation of compounds like acrylamide need to be considered in the course of reformulating the traditional products. For example, addition of wholegrains escalates acrylamide content in cereal-based foods in comparison the one's formulated with refined grains, as more asparagine is present in wholemeal flour containing bakery products (Mesías *et al.,* 2019; Mesias *et al.,* 2022)

In this chapter, recent formulations for cereal-based products have been mentioned in addition to the technologies developed for the processing of these products and improving their sensory attributes viz texture, volume, etc.

Cereal-based products

Snacks

Amaranth-based multigrain snacks bar

The growing consumer demands for high quality and healthy products has led to research aimed at including multigrain flours in snacks. This may remarkably enhance the nutritional profile while utilizing their health promoting properties. The consumer requirements can be met along with healthy nutrients through a snack bar. Olagunju *et al.* (2022) developed amaranth-based multigrain snacks bar using pre-fermented amaranth, acha, and pearl millet flours. As compared to locally available oats snack bar, this product provides better nutrition (supplies more than 25% RDA of protein and contains notable mineral elements) and antioxidant potential and has a low glycemic response (Olagunju *et al.,* 2022).

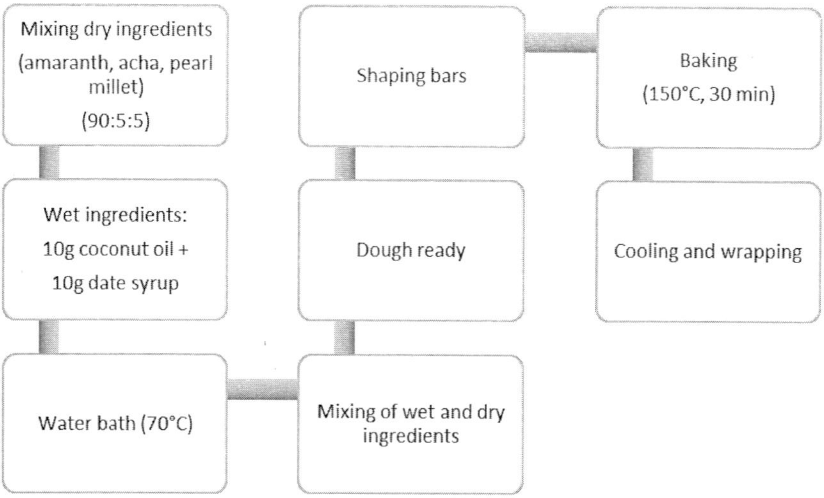

Process flow for preparation of Amaranth-based multigrain snacks bar (Olagunju et al., 2022).

Multigrain extruded snack

Due to availability, acceptability, and ease of preparation, extruded snacks are a common component of most people's diets. Generally, grains high in starch such as wheat, rice or corn are used to prepare the extruded snacks. Extrusion cooking is a high temperature, short time, commercial process for producing products that are nutrient-rich (Rodriguez-Miranda et al. 2011, Jing and Chi 2013).

Karun *et al.* (2023) developed a ready to eat extruded snack combining pearl millet, corn, rice, green gram and cowpea bean in the ratio of 30:30:25:8:7 using a high shear counter rotating twin screw extruder. The snack is extruded at a barrel temperature of 120 °C with a screw speed of 350 rpm and 10% (moisture content) with desirability. In order to obtain maximum expansion ratio, the particle size of 0.500 mm is maintained. The particle size is decided by considering factors such as moisture, fat and protein content of each grain. The prepared formulations are fed in the extruder by a calibrated feeder. The extrudates then pass through a 2 mm die where a rotating cutter is present. Finally, the extrudates are collected and air dried.

The multigrain extruded snacks can be considered a plant-based protein snack. To determine how extrudates' sensory and functional qualities are retained by correct packaging and seasoning in stability studies of the finished product, more research is needed (Karun *et al.*, 2023).

Millet based pizza based

In India, bakery goods production has witnessed a two-fold rise in recent years. The bakery business has a notable presence in the Indian industrial market alongside traditional bakery goods including biscuits, cakes, cookies, etc. (Kamliya and Rema, 2003). Pizza was one of the most often bought commodities in retail food stores, and despite the processed food business evolving, it managed to keep its market share and even grow in popularity. Pizza has markets in America, Europe, and other continents and has accelerated the trend towards convenience meals and global cuisine. As a result, pizza manufacturing has expanded significantly and is anticipated to do so further over the course of the next ten years due to the expanding global population (Preeti and Goyal, 2011).

Owning to the above reasons, pizza enrichment can lead to nutritional and health benefits of the population. Millet-based pizza base was developed by combining proso millet flour and refined flour in ratio of 70:30 and baking at a temperature of 200°C. The sensory profile of this product (crust color, taste, texture, overall acceptability) was remarkably similar to the wheat based pizza base (Bhavya *et al.*, 2020).

Fish-cereal based snack

Extrusion process has also been used for production of snacks by incorporating fish flesh or fish powder in starch-based stuff, which was accepted by consumers (Gogoi et al. 1996; Suknark et al. 2001). Single-screw and twin-screw extruders can be used for such purposes. Moreover, the nutritional quality of these extrudates can be enhanced by integrating fish hydrolysates and cereals (Choudhury et al. 1998). Rhee et al. (2004) effectively created a snack food by extruding defatted soya flour, maize and catfish mince. The fish-based products shall help fight malnutrition in developing countries by supplying nutrition and balanced diet (Venugopal and Shahidi 1995).

According to Shaviklo *et al.* (2011) when the acceptability of corn snacks fortified with varying amounts of fish protein powder is compared, snacks fortified with 9% fish protein powder demonstrated significantly lower liking in terms of odour, texture, flavour, and overall acceptability, whereas snacks with 3%, 5%, and 7% fish protein powder demonstrated similar sensory attributes with better acceptability (Singh *et al.*, 2014).

Fermented foods

In sub-Saharan Africa, traditional fermented foods make up the largest percentage of the diet. Majority of these foods are made from cereal grains (maize, sorghum, and millet), root tubers, legumes, and various seeds. These foods are still produced using traditional techniques which are being followed over generations (Sanni, 1993). Nevertheless, it has been observed that lactic acid bacteria govern the fermentation process of foods like fufu and ogi (Kostinek *et al.*, 2007; Oguntoyinbo, 2007; Sanni *et al.*, 2013). Some of the other common bacteria involved are Streptococcus, Leuconostoc, Pediococcus, Bacillus and Micrococcus (Chavan and Kadam, 1989). Although *Zygosaccharomyces*, *Geotrichum*, and *Torulopsis* have also been detected in various foods, Saccharomyces and Candida yeasts are the most commonly found yeast genera (Gotcheva *et al.*, 2000).

Yosa

Yosa is a snack prepared out of oat bran pudding that has been boiled in water and fermented with LAB (Lactic acid bacteria) and Bifidobacteria. Following fermentation, fruit jam and sucrose or fructose is added to flavour it (Salminen & Von Wright, 1998). This product is primarily consumed in Finland and other Scandinavian nations. It has a sensory profile close to yoghurt anyhow it does not contain milk or other animal products (Toufeili *et al.*, 1997). Hence, it is low-fat, free from lactose, comprises of b-glucan and is fit for vegans (Bioferme, 1999). The oat fibre and probiotic LAB present in Yosa may maintain and enhance the consumer's intestinal balance and therefore is a healthy incorporation in the diet (Toufeili *et al.*, 1997). The b-glucan found in oat fibre is also a good source of lowering blood cholesterol, which lowers the risk of heart disease (Bioferme, 1999; Blandino *et al.*, 2003).

Boza

Boza, a Bulgarian thick, sour or sweet, fermented cereal-based beverage is consumed in some parts of Turkey, Albania and Romania. It is often consumed with cinnamon and roasted chickpeas with molasses (Basaran, 1999). Boza is prepared by fermenting several types of grain (such as wheat, corn, rice, barley and oats) with lactic acid bacteria and yeast that are present naturally. The sensory attributes, nutritional value, and digestibility of food items are all improved by yeast and the lactic acid fermentation (Steinkraus, 1994).

Some of the other beverages similar to Boza are sekete, Kenyan busaa, Nigerian ogi, pito, South African kaffir beer, Egyptian bouza, and Turkmenistan and Krim busa (Sanni, 1992; Steinkraus, 1996; Beuchat, 1995; Adegoke, 1995; Pederson, 1979; Hayta *et al.*, 2001)

Other products

Rice flake-milk pudding instant mix

Milk pudding is a complex product mostly prepared by cooking cereals or legumes (powder or granulated) in milk with sweeteners. The puddings may be semi-liquid or thick gel-like which can be sliced. It is a dessert popular in celebrations and festivals (Chandan and Kilara, 2015). It is commonly prepared in most households but the preparation is a time-consuming and exhausting task. Also, the quality of product varies with the cook (Unnikrishnan *et al.*, 2003). Hence, milk pudding offers the dairy processing industry great opportunities for product diversification and market penetration, especially as a convenient dry mix with added value and extended shelf life. The instant mix can be prepared by dry blending ingredients such as rice flakes (pre-gelatinized sheets of rice batter cut into small flakes), sugar and milk powder in standardized proportions (Unnikrishnan *et al.*, 2003).

Although, the dry blend when reconstituted for consumption lacks the quality in terms of texture and flavour achieved from traditional (slow cooking) procedures. To overcome this, many alternate approaches have been developed including the dry-crystallization technology (Jose *et al.*, 2022). The dry-crystallization technology has been discussed in the further sections.

Development of cereal product quality

Sourdough

Cereal bran is the most common source of dietary fibre in bakery products. The use of hulled barley bran or hulled oat bran is also gaining popularity owing to the high soluble dietary fibre content, especially their b-glucan content. Nonetheless, adding cereal bran in amounts which can provide health benefits, lead to low quality of bread. The following changes in bread quality were observed by Seibel (1983) on addition of dietary fibre:

- Higher dough yield
- More moist and short dough
- Low fermentation tolerance
- Less bread volume
- Tense and non-elastic crumb
- Changes in flavour based on fibre and bread type

One of the ways to enhance high-fibre wheat breads quality is fermentation or pre-soaking of wheat bran prior to addition to dough. It has been suggested that pre-soaking wheat bran for an hour may increase loaf volume (Lai, Davis, & Hoseney, 1989; Lai, Hoseney, & Davis, 1989). In wheat bread containing 12% bran, hydrating the bran for 15 minutes in extra water at 10 or 97 °C before adding it to the dough increased loaf volume and enhanced bread quality (Nelles, Randall, & Taylor, 1998). Nelles *et al.* (1998) suggested three probable causes for the observed improvement: better lipoxygenase activation, washing out of free reduced-glutathione, and improved hydration of all brown flour components. According to Salmenkallio-Marttila *et al.* (2001) longer soaking times (4–16 h) or intermediate temperatures (25–30 °C) does not increase loaf volume. Moreover, various microbes are present in the bran layer of cereals, for example Bacillus subtilis, hence, care must be taken when soaking the bran in environments that encourage the growth of bacteria in order to minimise potential microbiological concerns. Furthermore, other safety concerns of whole grain cereals or fibre products should always be taken care of, such as the absence of biotic or xenobiotic compounds.

According to Salmenkallio-Marttila *et al.* (2001), pre-fermenting bran with yeast, or more specifically, with yeast and lactic acid bacteria, increases loaf volume and crumb softness during storage. Sourdough technique can be used successfully in high-fibre baking as well. According to Boskov-Hansen et al. (2002), the indigenous enzymes of flour, particularly the amylases and proteases, are thought to be responsible for the efficacy of fermentation. The pH of the dough is lowered by the acids created during fermentation, which has an impact on the properties of the gluten and the activity of the enzymes. Based on wheat type and germination status, the pH optima of carbohydrate-degrading enzymes, such as amylase, pentosanase, or b-glucanase, varies significantly (pH 3.6-5.6) (Fox & Mulvihill, 1982). Flour's gluten-associated proteinases are typically active when the pH is below 4. Slower fall in pH allows further starch degradation of the spontaneously fermented dough whereas a quick fall may inhibit amylolytic activity. The structure of the gelatinized starch granules is altered by starch modifications during fermentation (Eynard, Guerrieri, & Cerletti, 1995; Siljestrom *et al.,* 1988; Katina *et al.,* 2015).

Enzymes:

The quality and processing attributes of grain raw materials are influenced by endogenous enzymes. With the exception of malting, endogenous enzyme effects have often been minimised because it has been widely believed that they affect grain processing qualities. Industrial baking has for years utilised microbial enzymes (Hammer, 1991; Kulp, 1993; Fox and Mulwhill, 1982;

Linko and Linko, 1986). The benefits of adding enzymes as processing aids in baking include greater bread volume, enhanced crumb structure, and longer shelf life in addition to improved dough handling characteristics.

Due to the great specificity of individual enzymes, the sequence of modifications at various structural levels is typically triggered by the cleavage of particular linkages in dough biopolymers. This leads to the improved product qualities. The majority of the research on the application of enzymes in baking focuses on enhancing the volume, texture, and shelf life of bread. The use of bacterial and fungal α-amylases to enhance flour has a long history. External sources of α-amylases include barley and wheat malt (Poutanen, 1997).

The difference in dough rheology brought about by enzymes is the significant contributor to the enhanced bread volume. The dough can expand further as a result of softening, increasing oven spring (Cauvain and Chamberlain, 1988). Due to its altered composition and enhanced stability of the liquid film at the gas-liquid interface, the dough's ability to retain gas may also be improved. The manufacturing of frozen doughs depends heavily on the stabilisation of gas cells, which is a major goal of novel enzyme applications.

Enzymes can also help in maintaining the biscuit and bread quality when chemical additives are substituted. Hemicellulases, when used along with ascorbic acid, acts as a substitute to bromate, a bleaching agent banned in USA. α-Amylase-based bromate substitutes are also available. With the help of protease-based enzymes, sodium metasulphite can be replaced in biscuits (Sprossler, 1996). Moreover, the low amounts of oligosaccharides found in cereals can be increased using enzymes. The creation of prebiotic meals for the specific promotion of health-beneficial bacteria in the colon is of significant interest. The health benefits of oligosaccharides are now the subject of much research. It has been noted that arabino-xylo-oligosaccharides preferentially promote the development of Bifidobacteria (Yamada *et al.*, 1993), and using xylanolytic enzymes for their production may be taken into consideration (Poutanen, 1997).

DIC (Instant Controlled Pressure Drop):

When processing meals made from cereal and legumes, hydrothermal processes like steaming, boiling, and blanching account for a huge portion of the unit operations, particularly during production of groats, flakes, and cereal snacks. The DIC (Détente Instantanée Contrôlée) technique is a type of hydrothermal process that causes the food material to expand on its own as a result of the mechanical effect produced by the abrupt change from a high steam pressure (about 8.9 x 10^2 kPa) to a vacuum pressure of nearly 1 kPa (Iguedjtal *et al.,*2008). This process enhances product quality (biological and microbial decontamination, elimination of anti-nutritional compounds and conservation of nutritional value) as the expansion, cooking and texturizing operations occur in a very less amount of time (Haddad *et al.,* 2001; Mounir *et al.,* 2011). The effect of DIC on product quality can be controlled by altering process parameters like steam and vacuum pressure, moisture content of raw material and process time (Abdulla *et al.,* 2009). Hence, this technique can be useful for developing healthy ready-to-eat cereal based and legume-based snacks (Yağcı & Evci, 2015)

Dry-crystallization method:

Many instant mixes have been developed for the cereal-based milk puddings but the reconstituted products do not match the sensory profile of the traditionally cooked product. Dry-crystallization technology has been standardized as an alternate method by Unnikrishnan et al. (2003) by developing a mix that could get around these issues and win the reconstituted product the desired level of consumer acceptability.

During the dry-crystallization process, a mixture of milk, pre-gelatinized rice flakes, and sugar is continuously stirred and scraped in a steam jacketed kettle till the desired dry crystal form is attained. It takes very less time (10-12 min) to reconstitute the mix by cooking it in milk. The manual stirring required in the prior versions of dry-crystallization method was time-consuming and arduous. Hence, a specially constructed, mechanically stirred, batch-mode equipment was created for the standardised process of dry crystallisation in order to reduce the labor -intensive nature of the traditional preparation method (Panditrao *et al.,* 2021).

This technology was developed for dry-crystallized mix of rice flake-milk pudding and has an adequate opportunity of being used in industrial processes. Additionally, the dry-crystallization method used to create a convenient rice pudding mix has the potential to be of interest for additional variations of the product category around the world (Jose *et al.,* 2022).

References

1. Abdulla G, Belghit A. & Allaf K. (2009). Impact of controlled pressure drop treatment on moisture adsorption isotherm of cork granules. Drying Technology, 27, 237–247.

2. Adegoke GO, Nwaigwe RN, Oguntimein GB (1995) J Food Sci Technol India 32:516–518

3. Alvarez-Jubete, L., Arendt, E.K., and Gallagher, E. 2010. Nutritive value of pseudocereals and their increasing use as functional gluten free ingredients. Trends in Food Science and Technology 21: 106–113.

4. Angioloni, A., and Collar, C. 2011. Nutritional and functional added value of oat, Kamut®, spelt, rye and buckwheat versus common wheat in breadmaking. Journal of the Science of Food and Agriculture 91: 1283–1292

5. Angioloni, A., and Collar, C. 2012. Suitability of oat, millet and sorghum in breadmaking. Food and Bioprocess Technology DOI: 10.1007/s11947-012-0786-9.

6. Arendt, E.A., and Dal Bello, F. 2008. Gluten-Free Cereal Products and Beverages. Academic Press, Elsevier. Burlington, MA.

7. Basaran P (1999) Food Technol 53:60–66

8. Beuchat LR (1995) Indigenous fermented foods. In: Reed G, Nago-Dawithana TW (eds.) Biotechnology: enzymes, biomass, food and feed. Wiley, Weinheim, pp 505–590

9. Bhavya, P., Devi, S. S., Srinath, P. J., & Sreedevi, P. (2020). Development sensory evaluation of millet based pizza base. *IJCS*, *8*(4), 3502-3504.

10. Bioferme Oy. (1999). Yosa-The functional oat snack. Available: http://www.bioferme.fi.

11. Blandino, A., Al-Aseeri, M. E., Pandiella, S. S., Cantero, D., & Webb, C. (2003). Cereal-based fermented foods and beverages. *Food research international*, *36*(6), 527-543.

12. Boskov-Hansen, H., Andersen, M. F., Nielsen, L. M., Back-Knudsen, K.-E., Meyer, A. S., Christensen, L. P., & Hansen, A°. (2002). Changes in dietary fibre, phenolic acids and activity of endogenous enzymes during rye bread making. European Food Research and Technology, 214, 33–42.

13. Cauvain, S.P. and Chamberlain, N. (1988) 'The Bread Improving Effect of Fungal α-Amylase' in J. Cereal Sci 8(3). 239-248

14. Chavan JK, Kadam SS. (1989) Nutritional improvement of cereals by fermentation. Crit Rev Food Sci Nutr; 27:253–7.

15. Choudhury GS, Gogoi BK, Oswalt AJ (1998) Twin screw extrusion of pink salmon muscle and rice flour blends: effects of kneading elements. J Aquat Food Prod Tech 7(2):69–91

16. G. Panditrao, M.R. Ravindra, N. Jose, B.S. Nath, M. Sivaram, Process optimisation for the mechanised dry crystallisation of rice flake–milk pudding mix, Int. J. Dairy Technol. (2021), doi:10.1111/1471-0307.12834.

17. Eynard, L., Guerrieri, N., & Cerletti, P. (1995). Modifications of starch during baking: Studies through reactivity with amyloglucosidase. Cereal Chemistry, 72, 594–597.

18. Flander, L., Salmenkallio-Marttila, M., Suortti, T., and Autio, K. 2007. Optimization of ingredients and baking process for improved wholemeal oat bread quality. LWT—Food Science and Technology 40: 860–870.

19. Fox, P. and Mulwhill, D. (1982) 'Enzymes in Wheat, Flour and Bread' in Advances in Cereal Science and Technology Vol. 7(3) (Pomeranz, Y.. ed.), pp. 107-156, American Association of Cereal Chemists, St Paul, Mh. USA

20. Fox, P., & Mulvihill, D. (1982). Enzymes in wheat, flour, and bread In Advances in cereal science and technology, Vol. 5. St. Paul: American Association of Cereal Chemistry pp. 107–156.

21. Gogoi BK, Oswalt AJ, Choudhury GS (1996) Reverse screw element (s) and feed composition effects during twin-screw extrusion of rice flour and fish muscle blends. J Food Sci 61:590–595

22. Guiné, R. D. P. F., & dos Reis Correia, P. M. (Eds.). (2013). *Engineering aspects of cereal and cereal-based products*. CRC Press.

23. Haddad, J., Louka, N., Gadouleau, M., Juhel, F. & Allaf, K. (2001). Application of the new process of controlled instantaneous pressure drop (DIC) to the drying/texturization of fish: effects on physicochemical characteristics of the final product. Sciences Des Aliments, 21, 481–498.

24. Hamer, R.J. (1991) 'Enzymes in the Baking Industry' in Enzymes in Food Processing (Tucker. C.A. and Woods. L.F.J., eds), pp 168-193, Blackie

25. Hayta, M. E. H. M. E. T., Alpaslan, M., & Köse, E. (2001). The effect of fermentation on viscosity and protein solubility of Boza, a traditional cereal-based fermented Turkish beverage. *European Food Research and Technology*, *213*, 335-337.

26. Iguedjtal, T., Louka, N. & Allaf, K. (2008). Sorption isotherms of potato slices dried and texturized by controlled sudden decompression. Journal of Food Engineering, 85, 180–190.

27. Jing Y, Chi YJ (2013) Effects of twin-screw extrusion on soluble dietary fibre and physicochemical properties of soybean residue. Food Chem 138(2–3):884–889

28. Jose, N., Ravindra, M. R., & Deshmukh, G. P. (2022). Effect of dry-crystallization method on the engineering properties of an instant mix for rice flake-milk pudding. *Measurement: Food*, *7*, 100044.

29. Kamaliya KB, Rema S. Nutritional modification of the bakery products. Processed Food Industries. 2003; 7(1):23-27.

30. Karun, G., Sukumar, A., Nagamaniammai, G., & Preetha, R. (2023). Development of multigrain ready-to-eat extruded snack and process parameter optimization using response surface methodology. *Journal of Food Science and Technology*, *60*(3), 947-957.

31. Katina, K., Arendt, E., Liukkonen, K. H., Autio, K., Flander, L., & Poutanen, K. (2005). Potential of sourdough for healthier cereal products. *Trends in Food Science & Technology*, *16*(1-3), 104-112.

32. Kostinek, M., Specht, I., Edward, V.A., Pinto, C., Egounlety, M., Sossa, C., Mbugua, S., Dorti, C., Thonart, P., Taljaard, L., Mengu, M., Franz, C.M.A.P., Holzapfel, W.H. (2007). Characterization and biochemical properties of predominant lactic acid bacteria from fermenting cassava for selection as starter cultures. International Journal of Food Microbiology 114:342–351.

33. Kulp, K. (1993) 'Enzymes as Dough Improvers' m Advances in Baking Technology (Kamel, B.S. and Stauffer, C.E.. eds). pp. 153-l 78. Blackie Academic & Professional

34. Lai, C., Davis, A., & Hoseney, R. (1989). Production of whole wheat bread with good loaf volume. Cereal Chemistry, 66, 224–227.

35. Lai, C., Hoseney, R., & Davis, A. (1989). Effects of wheat bran in breadmaking. Cereal Chemistry, 66, 217–219.

36. Linko, P. and Linko, Y-Y. (1986) 'Enzymes in Baking' in Chemistry and Physics in Baking (Blanchard, J., Frazier, P. and Galliard, T., eds), pp. 105-116, The Royal Society of Chemistry, Cambridge, UK

37. Lorenz, K. J., & Kulp, K. (Eds.). (1991). *Handbook of cereal science and technology* (p. 815). New York: Marcel Dekker.

38. Mesías M, Sáez-Escudero L, Morales FJ, Delgado-Andrade C: Reassessment of acrylamide content in breakfast cereals. Evolution of the Spanish market from 2006 to 2018. Food Control 2019, 105:94-101.

39. Mesias, M., Delgado-Andrade, C., & Morales, F. J. (2022). An updated view of acrylamide in cereal products. *Current Opinion in Food Science*, *46*, 100847.

40. Mounir S, Besombes C, Al-Bitar N. & Allaf, K. (2011). Study of instant controlled pressure drop DIC treatment in manufacturing snack and expanded granule powder of apple and onion. Drying Technology, 29, 331–341.

41. Nelles, E., Randall, P., & Taylor, J. (1998). Improvement of brown bread quality by prehydration treatment and cultivar selection of bran. Cereal Chemistry, 75, 536–540.

42. Oguntoyinbo, F.A. (2007). Identification and functional properties of dominant lactic acid bacteria isolated at different stages of solid state fermentation of cassava during Gari production. World Journal of Microbiology Biotechnology 23:1425–1432.

43. Olagunju, A. I., Arigbede, T. I., Makanjuola, S. A., & Oyebode, E. T. (2022). Nutritional compositions, bioactive properties, and in-vivo glycemic indices of amaranth-based optimized multigrain snack bar products. *Measurement: Food*, *7*, 100039.

44. Pederson CS (1979) Microbiology of Food Fermentations, 2nd edn. AVI, Westport, pp 1–24

45. Preeti Singh, Goyal GK. Functionality of pizza ingredients, British Food Journal. 2011; 113(11):1322- 1338.

46. R.C. Chandan, A. Kilara, Puddings and dairy-based desserts, Dairy Process. Qual. Assurance 397 (2015), doi:10.1002/9780813804033.ch17.

47. Rhee KS, Kim ES, Kim BK, Jung BM, Rhee C (2004) Extrusion of minced catfish with corn and defatted soya flours for snack foods. J Food Process Pres 28:288–301

48. Rodriguez-Miranda J, Ruiz-Lopez II, Herman-Lara E, Martinez-Sanchez CE, Delgado-Licon E, Vivar-Vera MA (2011) Development of extruded snacks using taro (Colocasia esculenta) and nixtamalized maize (Zea mays) flour blends. LWT- Food Sci Technol 44(3):673–680

49. Rosell, C.M., and Marco, C. 2008. Rice. In Gluten Free Cereal Products and Beverages. Eds. E.K. Arendt and F. dal Bello. Elsevier Science. UK. 81–100.

50. Rosell, C.M., Cortez, G., and Repo-Carrasco, R. 2009. Breadmaking use of Andean crops quinoa, kaniwa, kiwicha, and tarwi. Cereal Chemistry 86: 386–392

51. Salmenkallio-Marttila, M., Katina, K., & Autio, K. (2001). Effect of bran fermentation on quality and microstructure of high-fibre wheat bread. Cereal Chemistry, 78, 429–435.

52. Salminen, S., & Von Wright, A. (1998). Safety of probiotic bacteria: Current perspectives. In T. Mattila-Sandholm, & T. Kauppila (Eds.), Functional food research in Europe (pp. 105–106). Finland: Julkaisija-Utgivare.

53. Salovaara, H., Kanerva, P., Kaukinen, K., and Sontag-Strohm, T. 2011. Oats-an overview from coeliac disease point of view. In The Science of Gluten-Free Foods and Beverages. Proceedings of the First International Conference of Gluten Free Cereal Products and Beverages. Eds. E.K. Arendt, F. Dal Bello. AACC Press, in press.

54. Sanni AI (1992) Int J Food Microbiol 18:85–95

55. Sanni, A., Franz, C., Schillinger, U., Huch, M., Guigas, C., & Holzapfel, W. (2013). Characterization and technological properties of lactic acid bacteria in the production of "Sorghurt," a cereal-based product. *Food Biotechnology*, *27*(2), 178-198.

56. Sanni, A.I. (1993). The need for process optimization of African fermented foods and beverages. International Journal of Food Microbiology 18:85–95.

57. Seibel, W. (1983). Anreching von Brot und Backverhalten. Getreide Mehl und Brot, 12, 377–379.

58. Shaviklo GR, Thorkelsson G, Rafipour F, Sigurgisladottir S (2011) Quality and storage stability of extruded puffed corn-fish snacks during 6-month storage at ambient temperature. J Sci Food Agr 91 (5):886–893

59. Siljestro¨m, M., Bjo¨rck, I., Eliasson, A.-C., Lo¨nner, C., Nyman, M., & Asp, N. (1988). Effects of polysaccharides during baking and storage of bread—in vitro and in vivo studies. Cereal Chemistry, 65, 1–8.

60. Singh, R. R., Majumdar, R. K., & Venkateshwarlu, G. (2014). Optimum extrusion-cooking conditions for improving physical properties of fish-cereal based snacks by response surface methodology. *Journal of food science and technology*, *51*, 1827-1836.

61. Sprossler B. 1996. 'Entwcklung und Einsatzmogllchkeiten von Enzymen bei der Backwarenherstellung' in Getreide, Mehl Brot 50, 281-283

62. Steinkraus KH (1994) Food Res Int 27:259–267

63. Steinkraus KH (1996) Handbook of indigenous fermented foods, 2nd edn. Dekker, New York

64. Suknark K, Lee J, Eitenmiller RR, Phillips RD 2001. Stability of tocopherols and retinyl palmitate in snack extrudates. J Food Sci. 66:897-902

65. Taylor, J.R.N., Schober, T.J., and Bean, S.C. 2006. Novel food and non-food uses for sorghum and millets. Journal of Cereal Science 44: 251-271.

66. Toufeili, I., Olabi, A., Shadarevian, S., Abi Antoun, M., Zurayk, R., & Baalbaki, I. (1997). Relationships of selected wheat parameters to bulgur-making quality. Journal of Food Quality, 20, 211–224.

67. V. Unnikrishnan, M.K. Bhavadasan, M.K. Vedavati, B.S. Nath, A dry mix for convenient preparation of palada payasam, Indian Dairyman 55 (7) (2003) 70-74.

68. V. Unnikrishnan, M.K. Bhavadasan, M.K. Vedavati, B.S. Nath, A dry mix for convenient preparation of palada payasam, Indian Dairyman 55 (7) (2003) 70-74.

69. Venugopal V, Shahidi F (1995) Value added products from underutilized fish species. Crit Rev Food Sci Nutr 35(5):431-453

70. Yağcı S, & Evci T. Effect of instant controlled pressure drop process on some physicochemical and nutritional properties of snacks produced from chickpea and wheat. *International Journal of Food Science & Technology*, 2015; *50*(8), 1901-1910.

71. Yamada H, Itoh K, Morishita Y. and Tanlguchl. H. (1993) 'Advances in Cereal Chemistry and Technology in Japan Structure and Properties of Oligomers From Wheat Bran' in Cereal Foods World 38, 490-492

Chapter - 11

Preservation of Meat; Chilling, Freezing, use of Heat, Dehydration, Smoking, Modified Atmosphere Storage, Ionizing Radiation

Authors

Disha Chavan

M.Sc Scholar, Dr. Balasaheb Sawant Konkan Krishi Vidyapeeth, Dapoli, Maharashtra, India

Pravin Zine

Ph.D Scholar, Vasantrao Naik Marathwada Krishi Vidyapeeth, Parbhani, Maharashtra, India

Nikhil Kadam

Ph.D Scholar, Dr. Balasaheb Sawant Konkan Krishi Vidyapeeth, Dapoli, Maharashtra, India

Chapter - 11

Preservation of Meat; Chilling, Freezing, use of Heat, Dehydration, Smoking, Modified Atmosphere Storage, Ionizing Radiation

Disha Chavan, Pravin Zine and Nikhil Kadam

Abstract

Meat has been consumed as a food in our daily diet from prehistoric times is an excellent source of protein and micronutrient. Origin of meat processing begins from humans learned that the cooking of meat prolongs their shelf life. The basic objective of meat processing was to improve the taste of meat and preserve it for long time. During the preservation meat is subjected to different process likes the chilling, freezing, smoking, curing and used different preservation methods like canning, ionization treatment. The benefit of meat preservation was processed food are convenient to use and handle and improves the nutritional value of meat. Due to preservation, we can enjoy any meat product at any season of the year or in the scarcity period. Preservation industries also generate the employment in the nation.

Keywords: Meat, prehistoric, smoking, ionization treatment, scarcity.

Introduction

Meat started their decaying with in a hours or day when it kept untreated because microorganism grows rapidly in meat products and meat becoming infectious or toxic. To avoid this spoilages meat preservation was very importance. The principles of meat preservation are mainly associated with preventing or delaying microbial spoilage and chemical action, and avoiding weight loss and any change in taste or texture. Methods of preservation are based on control by low or high temperature, reducing available moisture (aw), or by the application of different types of chemical preservatives. In developed countries the shelf-life of meat and meat products is, in the main, controlled by refrigeration or by a combination of refrigeration other methods of preservation, such as MAP, curing or high pressure processing. Whereas other techniques result in some changes to colour, taste, and texture, refrigeration and high pressure are distinct in that their effect on appearance and taste is minimal (Edward & D. A. 2003).

Definition: Meat preservation helps to control spoilage by inhibiting the growth of microorganisms, slowing enzymatic activity, and preventing the oxidation of fatty acids that promote rancidity. (Cross *et al* 2022).

Why meat preservation is important:

1) Improves the shelf life of meat
2) Enhance the product quality
3) Delayed the products spoilages
4) Off season availability of the products.
5) To control food borne infections and intoxications
6) To reduce raw material losses

Methods of meat preservation

1) Chilling
2) Freezing
3) Use of heat
4) Dehydration
5) Smoking
6) Modified atmosphere storage
7) Ionizing radiation

Initially, food preservation was done in the home. Knowledge how to smoke, cure, dry, ferment, salt, pack in fats or oils, pickle and deep in alcohol were abilities elemental to survival. Now a days, these food preservation skills have become the domain of food manufacturers, though they are also experiencing a revival on an individual level.

1. Chilling of meat

1) Chilling method was commonly used for the preservation of meat for very short period.

2) Chilling/refrigeration retards the spoilage rate at temperature below the optimal range can inhibit the microbial growth enzymatic as well as chemical reactions.

3) Storage of fresh meat is done at a temperature of 2 to 5°C.

4) Carcasses are first hanged in chilled coolers at 15°C to remove their body heat, and are then passed on to holding coolers to 5°C.

5) Chilling was done by two methods:

   i) Immersion chilling

   ii) Air chilling

6) Generally, fresh meat remains in good condition for a period of 5-7 days.

7) Temperature is the most important factor influencing bacterial growth produce in the meat products.

8) At lower temperature red color is more stable because the rate of the pigment decreases.

9) At low temp the solubility of oxygen is higher and oxygen consuming reaction is slowed down.

2. Freezing

1) The rate of freezing time is important in maintaining meat quality.

2) Rapid freezing is superior; if meats are frozen by sharp freezing, large ice crystals form in the meat and rupture cell membranes. When this meat is thawed, much of the original moisture found in the meat is lost as sweep out.

3) Due to this reason cryogenic freezing or other rapid/ flash/ super freeze methods of freezing meats are used at the commercial level to maintain maximal product quality.

4) The freezing does not kill most of the microorganisms; they simply become dormant in the freezing state.

5) Under proper freezer storage of -18 °C temp beef can be stored for 6 to 12 months, lamb for 6 to 9 months, pork for 6 months, and sausage products for 2 months.

3. Use of heat

1) Preservation by using heat has a different method:-

   i) Pasteurization

   ii) Blanching

   iii) Canning

   iv) Sterilization

2) Pasteurization refers to moderate heating of meat products in the temperature range of 58-75°C, which is also the cooking temperature. Heat treatment extends the shelf life of meat. It is necessary that such products also need to be stored under refrigerated conditions.

3) Sterilization refers to the heating of meat at the temperatures above 100°C where all spoilage causing micro-organisms in meat are killed or their microbial cells are totally damaged. The exposure of meat to high temperatures imparts sulrydryl flavors in cans and also modifies texture.

4) Canning is the process of preservation by thermal sterilization of a product held in hermetically sealed containers. Canning preserves the

organoleptic attribute such as texture appearance, flavor, and taste of the meat products to a large extent. Canned meat products have a shelf life of at least two years at ambient temperature. Canning involves various steps, such as preparation of meat, precooking, filling, exhausting, seaming, thermal processing, cooling, and storage.

5) Blanching of meat is done for 1 to 15 minutes depending upon the pieces size cuts. Blanching is mainly done or removes the blood and impurities from the muscles and bones and thus achieving the less cloudy broth.

4. Dehydration

1) Removal of water from meat pieces, the water soluble nutrients making them unavailable to the microorganisms is the main aim of dehydration.

2) In dehydration water activity should be lowers considerably to prevent the growth of spoilage causing microbes.

3) The Sun drying of meat chunks as a means of preservation was practiced in back days but rehydration of such meat used to be limited. In mechanical drying process involves the passage of hot air with controlled humidity but here also there is difficulty in rehydration of that piece.

4) Freeze drying of meat is a acceptable process of dehydration, preservation due to better reconstitution properties, nutritive quality and greater acceptability of the meat products as compare to the traditional dehydration methods.

5. Smoking

1) Meat smoking is also known as an aid in the preservation for a long time to the meat product.

2) It is now well that smoke contains a large number of wood degradation products such as aldehydes, ketones, organic acids, phenols and many more.

3) Preservation of meat by smoke is also due to surface dehydration, lowering the surface pH and antioxidant property of smoke constituents.

4) Recently, many liquid smoke preparations are commercially available in the many countries. Liquid smoke is mainly prepared from hard wood where polycyclic-hydrocarbons are removed by the process of filtration. Application of liquid smoke on the product surface before cooking imparts it a smoky flavor, which is very much liked by the consumer.

6. **Modified atmosphere storage**

1) In modified atmosphere storage the packaging involves the removal of air (oxygen) from package and either gases (co2, nitrogen) is filled in the packs or without any gas, pack is sealed.

2) Types of MAP for fish products: 1. Gas packing 2. Vaccum packaging

3) Oxygen (aerobic environment) is required for many bacteria to grow. For this reason most meats and meats products are vacuum-packaged, which extends the storage life under refrigerated conditions up to a 100 days.

4) In addition, vacuum packaging minimizes the oxidation of unsaturated fatty acids and slows the development of rancid meat.

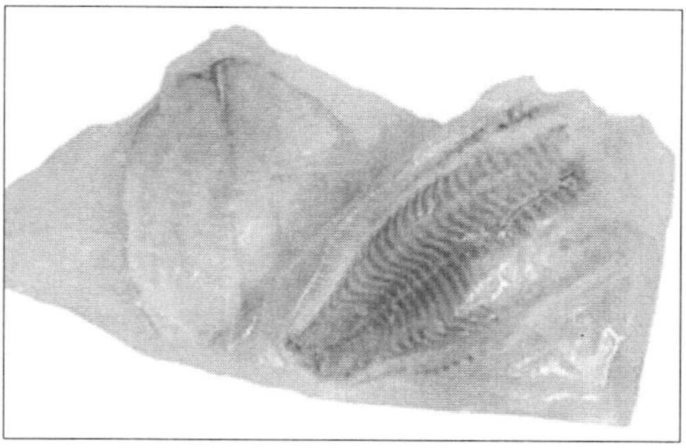

7. **Ionizing radiation**

1) Irradiation is a pasteurization method accomplished by exposing meat to doses of radiation.

2) Radurization is as effective as heat pasteurization in killing food-spoilage microorganisms. Irradiation of meat is done by exposing meat to high-energy ionizing radiation produced either by electron accelerators or by exposure to gamma-radiation-emitting substances such as cobalt-60 or cesium-137.

3) Irradiation of any food commodity up to an overall average dose of 10 Kgy causes no toxicological hazard hence toxicological testing of food so treated is no longer required, and irradiation of food upto 10 KGy introduce no special microbiological and nutritional problems

4) 4 Irradiated products are virtually identical in character to non-irradiated products, but they have significantly lower microbial contamination.

5) Irradiated fresh meat products require refrigeration and packaging to prevent spoilage, but the refrigerated storage life of these products is greatly improved.

Conclusion

With increasing demand for meat in the market preservation of meat is very important now days. Due to the different preservation techniques the meats provides very long shelf life and available the meat in any season of the year. The benefit of meat preservation was processed food are convenient to use and handle and improves the nutritional value of meat. Preservation industries also generate the employment in the nation. Due to preservation also avoid the post-harvest losses of meat.

References

1. Ledward, D. A. (2003). *MEAT | Preservation. Encyclopedia of Food Sciences and Nutrition, 3772–3777.* doi:10.1016/b0-12-227055-x/00752-5

2. Cross, H. Russell and Singh, R. Paul. "Meat processing". Encyclopedia Britannica, 30 Mar. 2022, https://www.britannica.com/technology/meat-processing. Accessed 6 May 2023.

3. Pal M, Devrani M (2018) Application of Various Techniques for Meat Preservation. J Exp Food Chem 4: 134. doi: 10.4172/2472-0542.1000134

Chapter - 12

The Proportion of Muscular tissue in Mutton and Poultry, Comminuted meat Products, Fresh Sausages, Cured Sausages, and Fermented Sausages

<u>Authors</u>

Neha Naijo Areekal

Ph.D. Scholar, Indian Institute of Technology, Tirupati, Andhra Pradesh, India

S. Menaka

M Tech. Dairy Chemistry, College of Food and Dairy Technology, Koduveli, Tamil Nadu, India

Yogeshwari. R

Assistant Professor, Sri Shakthi Institute of Engineering and Technology, Coimbatore, Tamil Nadu, India

Chapter - 12

The Proportion of Muscular tissue in Mutton and Poultry, Comminuted meat Products, Fresh Sausages, Cured Sausages, and Fermented Sausages

Neha Naijo Areekal, S. Menaka and Yogeshwari. R

Abstract

In today's era, the processing of meat products has expanded worldwide. A look at comminuted products reveals a wide range of fat and salt levels across categories; For instance, pork sausages contain between 11 and 30 grams of fat per 100 grams and between 0.9 and 2.5 grams of salt per 100 grams. Due to their influence on the flavor, which can serve as a point of differentiation between different brands of the same product, the wide range in fat and salt content makes it difficult to reduce their levels. One of the most delicious, wholesome, pleasurable, and practical meat products is sausage. Grinding beef, pork, chicken, or game meat and seasoning it with salt and other ingredients are the steps undertaken to make sausages, which are then stuffed into a casing or other container. Therefore, this chapter deals with the processing of meat products: the proportion of muscular tissue in mutton and poultry, comminuted meat products, fresh sausages, cured sausages, and fermented sausages.

Keywords: Mutton, poultry, comminuted meat products, sausages

Introduction

Muscle mass of livestock and its use to produce human food represent 35 to 60% of their body weight. The striated skeletal muscles attached to the backbone are involved in voluntary movements and facilitate locomotion and posture. Skeletal muscles exhibit a wide diversity of shapes, sizes, anatomical locations, and physiological functions. They are characterized by a composite appearance because, in addition to muscle fibers, they contain connective, adipose, vascular, and nervous tissues. Muscle fibers, intramuscular connective tissue, and intramuscular fat play key roles in the determination of meat quality. Concerning meat and aquatic products.

Creature development with respect to changes in connective tissue, the amount and level of cross-connecting in muscle connective tissue are largely connected with the basic strength of creature meat. Because it is impossible to completely control the properties of the muscle, a number of strategies are used to meet customer expectations. The first step is to sort the muscles by where they are on the carcass and what they do, which are typically linked to characteristics of connective tissue.

Muscle tissues can effectively be enriched with n-3 fatty acids, α-tocopherol, and Se, thus meeting the requirements for functional foods, whereas enrichment with CLA is less effective. In general, functional poultry meat has to be considered as a niche production due to the higher costs of these products, the limited application caused by changes in quality features, and due to limited proportion of consumers interested in purchasing the products.

Processed meat products are also a major source of fat and salt in the diet, reducing their fat and salt content could have a significant positive impact on public health in many countries due to their high consumption.

Meat products with health-improving additives are developed in response to rising consumer demand for high-quality meat products. In order to produce high-quality products, banger production requires careful meat selection. All sausage production formulas are based on meat, and any additions used in the production of sausages must be based on weight rather than the percentage (Abdolghafour and Saghir, 2014). The oldest type of meat products that have been processed are sausages, according to historical records. Salted or preserved by salting is what the Latin word salsus, from which the English word banger is derived, denotes (Mohan, 2014). Since many years ago, various and diverse cultures all over the world have used sausages as a food ingredients and processed meat products. Making sausages is a straightforward process that entails letting meat go through a number of carefully regulated chemical and structural changes. These are universally basic, but the variations rely on a variety of cooking and seasoning techniques to produce the desired distinctive flavours (Essien, 2003).

Proportion of muscle tissue in mutton

Goats are found worldwide today except for the extremely cold regions. The goats are present in the nations with broad Agricultural creation, despite the fact that their populace is ascending in the more extravagant nations. This could be mainly due to the bigotry of specific gatherings to cow's milk.

Although goat meat consumption is lower worldwide than beef consumption (Madruga and Bressan, 2011), goats unquestionably provide the majority of the world's red meat (Webb *et al.*, 2005), especially in developing nations.

Goat meat Quality

Goat meat is generally regarded as lean red meat with advantageous nutritional properties (Babiber *et al.*, 1990, Hogg and others, 1992; Webb and co., 2005). Compared to other meats of lamb or mutton, goat meat has a darker red color, a coarser texture, and a distinctively different flavor and aroma (Schönfeldt *et al.*, 1993 a,b; Casey and others, 2003). Sensory studies have shown that goat meat is different from lamb meat but not worse (Webb *et al.*, 2005). Additionally, goat meat and the products tend to be less succulent than mutton (Schönfeldt *et al.*, 1993b; Tshabalala *et al.*, 2003) prevalently because of their decreased fat substance.

Composition of muscle connective tissue

The proportion of connective tissue in muscle ranges from 1.5 to 15% of the dry matter, depending on the muscle's function. In a proteoglycan matrix, intramuscular connective tissue primarily consists of collagen and elastin proteins. Except for the muscles of the semitendinosus and latissimus dorsi, which have an elastin content of approximately 2% of the dry matter, muscle elastin content is approximately 0.1–0.2%. Collagen makes up the majority of this connective tissue, with concentrations ranging from 1.5% to 10% of dry weight.

Function of connective tissues

Skeletal muscle connective tissue shapes an organization that plays a dynamic role in muscle separation and improvement. In skeletal muscle, it serves as a structural support.

The adhesion of connective tissue to myofibers and fiber bundles allows for the sliding of fiber bundles and provides a scaffold for maintaining muscle structure during contraction. Connective tissue gives the phone-to-cell associations both between individual muscle cells and between muscle cells and adjoining little veins and nerves. It distributes the forces of muscular contractions in both muscle and tendon, acts as an elastic, stress-tolerant system, and provides mechanical strength and coherence. After being damaged, the connective tissue of the muscle takes part in cell growth and tissue regeneration.

The need to accommodate changes in the shape can be difficult as the muscle contracts are reflected in the organization of the muscle into fascicles separated by the perimysium, which allows fascicles to slide past one another. Practically, various muscles have different necessities for obliging the shear strains that essentially happen as the muscle contracts and changes shape, which makes sense of why the sums and conveyance of perimysium connective tissue shifts so broadly between practically various muscles.

Changes in connective tissue with cooking

Elastin is very heat stable, and its properties are generally unaffected by cooking. However, elastin content in the perimysium is generally low (0.1–0.2% of the total connective tissue) and its impact on meat texture is, therefore, considered as not significant. However, elastin may contribute to resistance to shear in the semitendinosus muscle, where elastin accounts for up to 40% of total connective tissue.

Even though connective tissue makes the most of a meat's texture when it is raw or undercooked, myofibers make the most of a meat's texture when it is cooked. The majority of the perimysium gels when cooked at temperatures between 60 and 70 °C. The temperature for primary changes tends to differ between studies: A few creators report hamburger perimysium i.e. Gelation at 70 °C or more, while others refer to 60-63 °C. Perimysium changed its thermal mechanical property at 50–60 °C, before the myofibers toughened, using isolated fibers and small muscle samples. Collagen thinning is primarily to blame for these changes. The Commitment of connective tissue to the meat surface decays as temperature Increments.

Comminuted meat products

Comminuted meat products typically contain a mixture of meat and non-meat ingredients, with the meat component typically being semi-lean meat from lower-value cuts of meat like trimmings and pork belly. The term "comminuted meat" is used to describe meat that has been cut, shredded, ground, or minced into small pieces, whether it is raw or pre-cooked. The quantity when pig meat, bird meat, or rabbit meat is used is 55%, and 62% when other meat (e.g., Beef) is used. For sausages, the minimum meat content is lower, as 42% is needed for pork Sausages, 26% for bird and/or rabbit sausages, and 30% for other meat sausages beef) is used.

Fat plays a role in sensory properties like taste and texture, which are crucial for a product's consumer acceptability. Additionally, it can be used as a substitute for lean meat in the product, lowering the formulation's cost. Due to the connection between excessive animal fat consumption and the risk of cardiovascular disease, consumers are concerned about the fat in comminuted products. The World Health Organization (WHO) recommends that less than 10% of calories should come from saturated fat and less than 30% from total fat (Flaherty *et al.*, 2012).

Processed meat fat and salt intake

Fat intake:

In the United States, the UK, and Ireland, fat intakes are currently higher than the WHO recommendation of less than 30% of total energy coming from fat whereas in Ireland, according to the NANS, fat accounts for 37% of the energy from daily consumption of food consumed by people aged from 18 to 64 and 36% people aged 65 and older. The information additionally showed that meats (entire and Processed), out of 16 food classes, provided 24% of the all-out fat admission making them the Greatest benefactor of fat to the eating routine (WHO).

Salt intake

It is critical to take note that salt isn't the main wellspring of sodium in comminuted meat Products since different fixings, for example, sodium triphosphate, sodium metabisulphite, sodium nitrate, and sodium bicarbonate are likewise added to sodium content. Consequently the relationship between salt (as NaCl) and sodium content shifts from one item to another.

Both the factors such as the type of comminuted meat product and the brand formulation within a product type influence the salt content of comminuted meat products. For instance, pepperoni, which has the highest average salt content, also varies in salt content per 100 grams, ranging from 1.4 grams to 4.4 grams. A single serving of comminuted meat can have a significant impact on one's salt intake; Using the example of pork sausages, one serving could supply anywhere between 0.54-1.380 g of salt (0.22-0.563 g sodium) per day, and as per recommendations of WHO, a sodium intake of 2g, that means that one portion of sausages would account for 11-28% of the daily intake (FSAI).

Proportion of muscle tissue in poultry

In addition to enzymes, sarcoplasm contains a variety of soluble, muscle-specific proteins with a variety of physiological functions. Myoglobin, one of these proteins, is present in muscle tissues in quantities that are proportional to their physiological or physical activities. Myoglobin has a higher oxygen affinity than hemoglobin. Large levels of myoglobin are found in the slowly contracting red muscles, where it primarily uses aerobic energy routes. The white muscle, on the other hand, relies heavily on anaerobic glycolysis for energy and is constituted mostly of a fast-contracting type of fibers despite having little myoglobin (Burke *et al.,* 1971; Peter *et al.,* 1972).

Parvalbumin, another highly soluble protein, was initially isolated and named from the white muscle of lower vertebrates by Pechere (1968). It is found in large concentrations in this muscle. It has an average molecular weight of 12000 and is an acidic calcium-binding protein. Although in much lesser amounts, parvalbumin has also been discovered in the muscle of higher vertebrates (Lehky *et al.,* 1974). It is only present in skeletal muscles (Baron *et al.,* 1975), with concentrations that are especially high in white, rapidly contracting muscles (Blum *et al.,* 1977). Parvalbumin acts as a relaxing factor in these muscles, removing calcium from the myofibrillar troponin C to promote faster relaxation before the sarcoplasmic reticulum recaptures calcium.

Function of poultry meat

The majority of poultry meat is made up of muscle, with varying amounts of connective tissues. Adipose tissue, bone, and cartilage are among its other components, just like some nervous and epithelial tissues. Meat's leanness and quality can be attributed to the properties and proportions of muscle and connective tissues, which are the most abundant tissues (Listrat *et al.,* 2016).

Biochemical and underlying attributes of muscle fiber can be autonomously controlled by natural and outward factors to accomplish creation effectiveness and further develop meat quality (Listrat *et al.,* 2016). The contractile, metabolic, and morphological characteristics of these muscle fibers are what set them apart (Astruc, 2014).

The morphology traits are classified by the total number of fibers (TNF), the cross-sectional area of fibers (CSAF), and fiber type composition (FTC). In general, muscle fiber types in adult skeletal muscle can be categorized as type I (slow oxidative), type IIA (fast oxide-glycolytic), type IIX, and IIB (fast glycolytic) (Schiaffino and Reggiani, 1996). All of these fiber types are observed in most mature mammalian skeletal striated muscles and their relative composition in the different muscles can determine the predominance of muscle's metabolic properties (Ozawa *et al.,* 2000; Ryu and Kim, 2005).

Sausage

A sausage is a charcuterie product that is mostly made of minced meat blended with additional ingredients such as spices and condiments. The preparation is then carried out by placing them in a gut, either natural or artificial, in the shape of a tube, and the ends are sealed. There are, however vegetarian sausages that are prepared from soy cheese or other ingredients, such as veggies. Different cultures around the world are able to produce sausages according to their local methods and the tastes of their consumers, allowing the length and size of the sausage to vary depending on the recipes (Gbaguidi *et al.,* 2020).

Types of sausages

Sausages generally differ according to the production process that is followed such as mincing of the seasoned meat, stuffing in casings, smoking, drying, fermenting, and heating. There are various sorts of sausages, and these variations often involve changing the type of meat, the temperature at which it is processed, the type of casings used, and the size of the casing particles (Savic, 1985). Table 1 shows the different types of sausages.

Fresh Sausages

Fresh meats are used to make fresh sausages; these meats are often not cooked, cured, smoked, fermented, or cured in any way. Before consumption, fresh sausages must be kept under refrigeration. Before serving, the customer heats them up himself (Abdolghafour and Saghir, 2014).

Fermented Sausage

The process of fermentation, one of the oldest ways to preserve meat, is used to make fermented sausages as lactic acid is produced during the fermentation process. Fermented sausages have a relatively longer shelf life than unfermented ones. Dry and semi-dry fermented sausages are the two categories:

- Dry - Salami, Pepperoni, Genoa
- Semi-dry - Summer sausage or cervelat, Lebanon bologna

In the manufacture, starter cultures-large quantities of cultured microorganisms-are used in a fermentation process together with curing agents, spices, and other additives. At the end of the procedure, products are produced as dried, cured commodities. In the past, acid bacteria found in the meat or new batches being inoculated with older batches have been used to create fermented sausages. The mixing stage, where the mixture is placed in casings was initially fermented and then dried where microflora is first introduced. Prior to cooking, several procedures allow for drying. By producing acid, fermentation inhibits the growth of pathogenic microorganisms (Essien, 2003).

Smoked Precooked

The majority of smoked precooked sausages are cured, non-fermented items; heating increases their shelf life by partially reducing their moisture content; they are often cooked completely before consumption (Abdolghafour and Saghir, 2014).

Emulsion-type sausages

Emulsion-type sausages are pre-cooked, usually smoked, and lightly cooked goods formed from minced and thoroughly homogenized cured meats, fatty tissue, water, and seasonings. These sausages are referred to as "scalded" in Europe because they have only been partially cooked (scalded; pasteurised) (Abdolghafour and Saghir, 2014).

Table 1 Sausage classification (Gbaguidi *et al.*, 2020)

Classification	**Characteristics**	**Examples**
	Raw sausages	
Fresh	Made from fresh comminuted, uncured, non-smoked meats. Must be refrigerated prior to heating by the consumer.	Breakfast sausage (USA), boerewors (South Africa), bratwurst (Germany), merguez (North Africa), siskonmakkara (Finland)
Fermented	Made from comminuted, cured or uncured, fermented, and often smoked meats. Not heat-processed.	
Semidry (quickly fermented)	Stuffed in medium- and large-diameter artificial casings. "Tangy" flavor produced by fermentation. Length of smoking and fermentation depends on type but rarely exceeds a few days. Improved stability of stored refrigerated	Variety of summer sausages, cervelats, mettwursts, Lebanon bologna (USA)
Dry (slowly fermented)		Different types of salamis, droë wors (South Africa, not fermented)
	Heat-processed sausages	
Smoked precooked	Mostly comminuted, cured, non-fermented. Final cooking before consumption	Chinese pork sausages, kielbasa
Emulsion type	Made from comminuted well-homogenized cured meats, fat, water and seasoning. Usually smoked, slightly cooked. Ready-to-eat product	Frankfurters, wieners, bologna, mortadella
Cooked	Made from previously comminuted cooked fresh or cured raw materials. Final cooking after stuffing. With or without smoking. Ready-to-eat product.	Liver sausage, Braunschweiger

Steps involved in the production of meat sausage

Raw material

High-quality meat that is raw should undergo microbiological testing. Prior to mixing, all ingredients should be accurately weighed in accordance with each unique recipe. Typically, the meat is either fresh or frozen. To ensure that the lean meat is completely free of ligament, bone, and cartilage fragments, it should be well-trimmed to a level of less than 10% non trimmable fat and connective tissue. Trimmed lean meat is also practically devoid of sinews and gristle. The choice of meat should be made with a good ability to bind water in mind. One of the best sources of high-quality meat sausages and ready-to-eat meat items is buffalo meat. (Kandeepan *et al.,* 2013; Abdolghafour and Saghir, 2014).

Grinding

The grinding of meat and fat ingredients has been widely used for many years and is still done today, primarily by small processors, especially in the creation of speciality products. Lean meat chunks are initially reduced using a 6-9 mm grinder plate, while fat trimmings or fatty tissues are first ground through a 3-6 mm grinder plate. It is best to grind the tripe and filler meats twice: once through a 3-4 mm grinder plate and once through a finer grinder plate. The meat plate is heated and the machine's capacity is increased by grinding through a coarser plate. It is thought that grinding through a finer plate will provide a product with greater binding and emulsifying qualities, especially in the case of bull flesh (Abdolghafour and Saghir, 2014). Following the addition of the curing salts, the mixture is mechanically mixed to ensure that all of the ingredients are evenly distributed. The curing procedure can be carried out either after the final cutting in the cutter with the other components and filling, i.e. before or during the smoking, or overnight in a chiller at 1-4°C. Because it aids in a better and more consistent size reduction in the cutter, comminution in the grinder which is then frequently followed by chopping during emulsion-type sausage procedures. Comminution is frequently not done with sufficient fineness. (Pearson, 1977).

Mincing

For simple protein extraction, the meat is chopped very finely after grinding. The purpose of proteins is to bind the water that surrounds fat droplets and maintain them distributed. Creating a banger emulsion essentially involves two steps. The lean meat is first placed on the cutter and chopped, whether it has already been ground or not. This is accomplished by adding one-third of the total volume of finely crushed ice or water, together with all of the curing salts (kitchen salt, nitrite), phosphate, and/or citrate for the batch at once.

The extraction of meat protein will increase with higher salt content in the water phase of the mixture, which is crucial for creating a stable emulsion.

Lean meat should be minced for a considerable amount of time, typically no less than 6 to 8 minutes. The remaining two thirds of the water are then added, followed by the addition of spices, fat trimmings, and additional fat meats. The batch is then chopped once more until it is finely chopped, and care is taken such that the temperature of the meat emulsion does not rise above 18°C (second phase). The decomposed and homogenized meats absorb all the water during this period. If sodium ascorbate is employed, it is also added just at the conclusion of the cutting process. More and more people are thus using these preparation techniques that offer one-phase along with simultaneous treatment of the meat and fat cutter. Before being added, the curing chemicals should be dissolved in a small amount of warm water in order to allow the polyphosphate's effect on actomyosin as well as the effects of salt and nitrite on the meat's water-binding qualities and colour to exert their effect. More salt soluble proteins are removed and the binding quality of the finished product is improved due to the high salt content and the lengthy cutting process. The use of phosphates is not required when using hot boned meat (Abdolghafour and Saghir, 2014).

Filling

Before adding the mixture to the casings, oxygen should be removed from it (using vacuum filling equipment) and the mixture's temperature shouldn't be higher than 2°C. The two types of casing that are most frequently used are natural (formed from the intestines of slaughter animals) and modified collagen or cellulose casings. There are also four undehydrated goods and synthetic casings. Air removal from the product improves color stability and the sausages' aesthetic appeal. Additionally, it lessens bacterial activity, inhibits proteolysis, and reduces fat oxidation. Vacuum filling allows sausages to have a longer shelf life (Essien, 2003).

Casings

Casings, commonly referred to as skins, are used in the production of sausages and serve primarily as containers for portioning. In general, they can be split into two categories: natural and artificial. Natural casings are made from sheep or pig intestines. After filling and cooking, most items made of natural casings have a curve when they are finished. Collagen, cellulose, and plastic materials are currently used to create artificial casings for a variety of uses. Collagen is taken out of animal connective tissue through a series of mechanical and chemical processes and used to make casings. Casings not only give the desired sausage shape, but they also extend the shelf life of the product by having strong seals and being highly moisture and oxygen resistant. Casings help to reduce product weight loss while cooking as a result (Essien, 2003).

Cooking

Cooking can be done in a variety of ways, including submerging in the cooking vat, hot showering in a smokehouse with shower nozzles, hot showering in separate hot water spray cabinets to which sausages are moved immediately after smoking, cooking by dry heat by raising the smokehouse temperature and giving only a final, quick hot water shower, cooking in small containers into which live steam is injected, etc. The schedules for cooking, however, differ greatly. The temperature is between 80-82°C while using water sprays and water in cooking vats may be between 73 to 76 degrees Celsius. The ideal end-point temperature for preserving the product's shelf life and achieving the appropriate organoleptic features is 68°C, however 65°C is often regarded as the minimum final internal sausage temperature. It normally takes 15 to 20 minutes to complete. By carefully examining the yield and quality of the sausage, the ideal cooking schedule should be created. Sausages are hot-showered after being cooked in vats to get rid of any remaining grease (Abdolghafour and Saghir, 2014).

Cooling and packaging

The sausage should be rinsed or submerged in water until it reaches an internal temperature of 38-40°F, at which point the casing should be removed and the sausage cooled to 4°F before packaging. Packaging is primarily used to protect meat and meat products from microbiological infection, physical alterations, and chemical alterations. Primary or secondary sausage packaging materials should be sufficient to provide a consumer with a respectable structural and visual presentation of the product. In order to increase shelf life, vacuum packing is used on frankfurters and prepared sliced sausages. The use of vacuum packing in cooked, refrigerated, and frozen products is growing in popularity. Vacuum packaging of sliced sausages is ideal for the sandwich and meal sectors in the case of packs that are sterilized and sanitized before use (Essien, 2003). In order to maintain the quality and extend the shelf life of meat and meat products during refrigeration, it is important to package them with the proper plastic film and laminates. Buffalo meat and meat products that have been pre-blended were vacuum-packed for improved refrigeration storage stability (Sahoo and Anjaneyulu, 1997).

Functional ingredients/additives in sausage manufacturing

Wet, semi-dry and dry ingredients

There are many different moist, semi-dry, and dry components that can be used to make sausages. Hog cutting fat, hog shoulder trim, pork back fat, liver, venison, rind, and mechanically recovered chicken or turkey flesh are some examples of meat constituents.

To create a seasoning mix, a variety of substances can be mixed together in various ratios. Dextrose, salt, herbs, sugar, spices, preservatives, onions, leek pieces, taste enhancers, antioxidants, stabilisers, hydrolyzed vegetable protein, yeast extract, parsley, pepper, and flavourings (natural, nature-identical, or artificial) are some examples of these ingredients.

As binders, numerous dry substances have been employed, including rusk, gelbind, potato starch, wheat flour, farina, cornflour, soya, and rice.The inclusion of tomato, egg, milk powder, milk protein, apricot, strawberry, beer, stilton cheese, and even various meat extenders and analogues might result in a diversity of flavour and possibly a different texture. Products made of protein from plants, such soy beans, have been utilised as meat substitutes and extenders. Vegetarian banger choices provide a lot of possibility for meat substitutes, extenders or analogues (Essien, 2003).

Antioxidants

One of the most common antioxidants used in sausage manufacture is ascorbic acid (E300). By reducing fat rancidity and colour change brought on by exposure to oxygen in the air, the intention is to increase the product's shelf life (Essien, 2003).

Preservatives

By inhibiting the development of bacteria that lead to food degradation and food poisoning, preservatives are used sparingly to extend the shelf life of food. Without preservatives, sausages would only last a very short time. To impart and maintain the product's colour, several preservatives are added. E221 and E223 are two of the most widely utilized (Essien, 2003). It took decades for nitrite to be used to preserve and cure meat. It is one among the earliest methods of food preservation. It has many advantages for banger manufacture, including increased food safety, longer shelf life, and good storage stability. Many of the popular cured sausage products on the market today include sodium nitrite. No other ingredient can replicate the distinct colours, textures, and flavours that curing provides. To obtain the distinctive flavour, colour, and stability of cured meat, certain curing chemicals are necessary. Microorganisms convert nitrate and nitrite into nitric oxide, which reacts with the flesh pigment (myoglobin) to provide the distinctive pink colour of cured beef. In contrast, nitrite works to prevent rancidity, stabilises the flavour of cured meat, and protects against the growth of germs that cause botulism (Mohan, 2014).

Flavour enhancers

By altering taste perception, a flavour enhancer supports a product's natural flavour. The production of sausages makes extensive use of monosodium glutamate (E621). The use of E621 has become less common as a result of customer concerns (Essien, 2003).

Flavourings, spice and herb extracts

The use of various herb and spice extracts to enhance flavour or as flavourings in sausage production is a result of advances in product development. Among the flavourings, spices, and herbs used in sausages are sage, parsley, coriander, paprika, rosemary, nutmeg, leek, apricot, cranberry, tomato, stilton, onions, and mint. These are packaged in the appropriate quantity to be added to the batch at the mixing point, either separately or as a seasoning mix. In terms of the variety of natural components and extracts chosen by consumers, sausage production has evolved, which in turn motivates the creation of new products. This interest is inextricably linked to consumers' desire to use fewer E-numbered additives in foods today (Essien, 2003).

Salt

In the process of manufacturing sausage, salt's potential to improve the flavour, preservation, water absorption, protein solubilization, and retention capacity of the meat has been effectively employed. Either directly or as a component of the seasoning mix that can be batch packed, salt is supplied as NaCl at a rate of around 0.5%. The amount of sodium in the finished product will be significantly influenced by the salt supplied at the mixing point as well as that from other ingredients such as rusk, gelbind, cure compound, and seasoning mix. If a low-sodium/low-salt product is required, calculate how much salt is contained in other compound ingredients before addition of extra salt. The product range is always being expanded in this direction of customer's preference I.E, low salt option while demanding a new product composition. To attain the desired functioning with less salt, this will necessitate significant adjustment of other additives (Essien, 2003).

Emulsifiers–stabilizers

Emulsifiers and stabilisers improve the close holding of oil, fat, and water in a mixture. Diphosphates (E450) and triphosphates (E451) are the stabilisers that are most frequently used for producing sausage (Essien, 2003).

Non-meat binders

Exudate formation cannot always be simple to produce in recipes with a high fat content and little lean meat. However, using other non-meat ingredients in these recipes that can set at heat to produce or improve the binding ingredients for a sausage producer is simple to do. These include whey powder, egg white, milk powder, soy protein isolate, and starches. Since they do not contribute to the recipe's meat content, it is obvious that good recommended functional levels of inclusions of these binders are needed (Essien, 2003).

Conclusion

The processing of meat products has witnessed tremendous growth in the poultry sector. Production of minced meat or long-term moist cooking processes requires cross-linked collagen to make meats more tender as it can strengthen the muscles. Furthermore, an assessment of the general benefit of functional poultry meat is not yet possible due to limited information on proven RDI levels of these substances and on probable interactions between the substances, which may result in a risk to human health.

The desire to get fresh meats, items that can be maintained in order to spread their use over time, has largely led to the processing of meats. Additionally, the industry has expanded its activity to include the production of poultry cold cuts in order to meet consumer demand. Freshly minced meat from a variety of meat species, including pork, cattle, chicken, fish, and buffalo, is used to make sausages. Like other cold meats, sausages give the body energy, lipids, and proteins with high biological value.

References

1. UK Food Standards Agency (2015). Labeling and Composition of Meat Products. Guidance Notes. September 2003. (Accessed on 17th September 2015).
2. Food Safety Authority Ireland. Salt Reduction Programme (SRP) – 2011 to 2012: Written Achievements and undertakings: submitted by the food industry to the Food Safety Authority of Ireland (FSAI).
3. World Health Organisation. (2014). Food-based dietary guidelines in the WHO European Region. (Accessed 16th October 2014).

4. O'Flaherty, M., Flores-Mateo, G., Nnoaham, K., Lloyd-Williams, F., Capewell, S., (2012) Potential cardiovascular mortality reductions with stricter food policies in the United Kingdom of Great Britain and Northern Ireland. Bull. *World Health Organ*. 90, 522-531.

5. Peter, J. B., Barnard, R. J., Edgerton, V. R., G, C. A., and Stempel, K. E., (1972). Metabolic Profiles of three fiber types of skeletal muscle in guinea pigs and rabbits. *Biochemistry,* 11: 2627-2633

6. Burke R.E, Levine D.N, Zajac F.E, Tsairis P. and Engel W.K. 1971. Mammalian motor units: physiological-histochemical correlation in three types in cat gastrocnemius. *Science*, 174: 709-712

7. Pechere J-F. (1968). Muscular parvalbumins as homologous proteins. *Comparative Biochemistry and Physiology*, 24: 289-295.

8. Lehky, P., Blum, H. E., Stein, E. A. and Fischer, E. H. (1974). Isolation and characterization of parvalbumins from the skeletal muscle of higher vertebrates. *Journal of Biological Chemistry*, 249: 4332-4334.

9. Baron G, Demaillie J. and Dutruge E. (1975). The distribution of parvalbumins in muscle and in other tissues. *FEBS Letters*, 56: 156-160.

10. Blum H.E, Lehky P, Kohler L, Stein E.A. and Fischer E.H. Comparative properties of vertebrate parvalbumins. *Journal of Biological Chemistry*, 1977. 252: 2834-283

11. Astruc T. (2014). Muscle structure and digestive enzyme bioaccessibility to intracellular compartments. *Food structures, digestion, and health*, 193-222.

12. Schiaffino S, and Reggiani C. (1996). Molecular diversity of myofibrillar proteins: gene regulation and functional significance. *Physiological reviews*, 76(2), 371-423.

13. Ryu, Y. C., and Kim, B. C. (2005). The relationship between muscle fiber characteristics, postmortem metabolic rate, and meat quality of pig longissimus dorsi muscle. *Meat science*, 71(2), 351-357.

14. Abdolghafour B, and Saghir A. 2014. Development in sausage production and practices-A review. *Journal of meat science and technology*, 2(3), 40-50.

15. Essien, E. (Ed.). (2003). *Sausage manufacture: Principles and practice*. Woodhead Publishing.

16. Gbaguidi M.D.D, Degnon R.G, Konfo T.R.C, Kpatinvoh B, and Baba-Moussa, F. (2020). Improvement in sausage manufacturing process and

potential use of selected aromatic plants as their bio preservatives in Benin. *GSC Biological and Pharmaceutical Sciences*, 11(3), 166-176.

17. Kandeepan, G., Mendiratta, S. K., Shukla, V., and Vishnuraj, M. R. (2013). Processing characteristics of buffalo meat-a review. *Journal of Meat Science and Technology*, 1(1), 01-11.

18. Mohan, A. (2014). Basics of sausage making, formulation, processing and safety anand. *UGA Extension*, 1-48.

19. Pearson, A. M., Love, J. D., and Shorland, F. B. (1977). "Warmed-over" flavor in meat, poultry, and fish. *Advances in food research*, 23, 1-74.

20. Sahoo, J., & Anjaneyulu, A. S. R. (1997). Quality improvement of ground buffalo meat by pre-blending with sodium ascorbate. *Meat Science*, 46(3), 237-247.

21. Savic IV. (1985). *Small-scale sausage production.* Rome, Italy: Publications Division, Food and Agriculture Organization of the United Nations.

Chapter - 13

Cleaning and Sanitation; Cleaning Procedures, Cleaning agents, Disinfection Techniques, Disinfectants for the Meat Industry, Cleaning and Disinfection Plans

Authors

Nivedhika Rai Vaid

Assistant Professor, Chandigarh College of Hospitality, CGC, Landran, Punjab, India

Pooja Bhatt

Assistant Professor, Chandigarh College of Hospitality, CGC, Landran, Punjab, India

Mamta

Ph.D. Scholar, Department of Food Science, Nutrition and Technology, CSK Himachal Pradesh Agricultural University, Palampur, Himachal Pradesh, India

Vaishnavi

Ph.D. Scholar, Department of Food Science, Nutrition and Technology, CSK Himachal Pradesh Agricultural University, Palampur, Himachal Pradesh, India

Chapter - 13

Cleaning and Sanitation; Cleaning Procedures, Cleaning agents, Disinfection Techniques, Disinfectants for the Meat Industry, Cleaning and Disinfection Plans

Nivedhika Rai Vaid, Pooja Bhatt, Mamta and Vaishnavi

Abstract

Cleaning, sanitation, and disinfection in preventing the spread of infections and maintaining a healthy environment. Effective cleaning procedures remove dirt, debris, and germs from surfaces, reducing the risk of contamination and providing a sanitary atmosphere. Sanitization aims to bring bacteria levels down to a safe level, often utilizing heat or chemicals. High-touch surfaces should be regularly cleaned to prevent the spread of germs. Disinfectants offer an additional layer of defense against disease-causing pathogens, effectively eliminating a wide range of bacteria, viruses, and fungi. They come in various forms and are particularly important in places where people gather, such as healthcare institutions and schools. These practices not only promote hygiene but also improve productivity, reduce absenteeism, and contribute to overall health. In conclusion, implementing proper cleaning, hygiene, and disinfection procedures is crucial in disease prevention, making environments safer and less susceptible to infectious illnesses.

Introduction

The cleaning and sanitation (C&S) of equipment, materials, and premises is done to prevent or mitigate the spread of foreign animal diseases (FADs) during an outbreak. This contributes to the stabilization of the food supply, the economy, and public health while also preserving the environment as part of a larger response. This standard operating procedure (SOP) gives direction on selecting and utilizing the best C&D techniques after a FAD epidemic in domestic cattle and poultry to State Animal Health Officials, APHIS workers, and Incident Management Teams (IMTs).

Programs for cleaning and sanitization are a crucial requirement to guarantee the adoption of good manufacturing practices (GMP). Cleaning won't be successful unless management completely supports it. The management's responsibility is to specify the necessary hygiene standards and to communicate these efficiently to personnel, typically using a thorough cleaning schedule and related task procedures. By providing the necessary resources, such as personnel training, equipment, and chemicals, management must show its commitment.

Cleaning schedules can be created to address daily, weekly, monthly, quarterly, six-monthly, and yearly tasks separately or all at once. Cleaning can either be observed by a hygiene team leader or equivalent as part of production processes, or it can be observed by the quality control function as part of a regular site hygiene audit. Verification of cleaning will be done on a regular basis by qualified individuals, such as the quality control manager or designate, and will involve a range of activities, such as record auditing, visual inspections of the premises, analysis of swabbing results, food product microbiological testing, customer complaint data, etc.

Definition

Cleaning Complete removal of food soil using appropriate detergent chemicals under recommended conditions from devices, work surfaces, workstations, and environment and so on.

Disinfection A process of destroying microorganisms, but usually not a bacterial spore, until their number is reduced to a safe level or reduces them to an acceptable level with respect to the defined purpose (a level that will not affect the safety or the food spoilage). Disinfection can approximately mean killing all active (vegetative) microbes but not the spores.

Disinfectants A chemical used to disinfect. These are usually chemical in nature and used in conjunction with detergents used to destroy, slow down growth and replication or removal of most microorganisms on surfaces, in or Facilities, appliances, accessories and equipment. Such products can be called germicides, bactericides or biocides.

Sanitation A chemical agent used for cleaning and disinfecting surfaces and equipment usually contains both detergents and disinfectants. To sanitize is to reduce the number of microbes to a safe acceptable level. It never recommended mixing cleaning chemicals on their own discretion.

Overview of the elimination of viruses

In order to stop the spread of the illness and prepare for restocking, C&S actions are carried out during a FAD epidemic as part of routine biosecurity operations (such as vehicle disinfection). The latter the cleaning and disinfecting of previously contaminated locations is also known as virus eradication. Same is the case with COVID-19 virus the transmission through surface contact of aerosol droplets from infected persons, followed by touching nose, eyes and mouth.

By lowering the number of dangerous microbes and stopping the spread of disease agents, C&S is a successful method for reducing the threat of animal diseases. For the purpose of removing, inactivating, reducing, or destroying harmful germs, C&S may involve the employment of physical, chemical, or biological methods. The physical qualities of the site and other elements, such as the environment, that may affect the efficacy of virus removal should be taken into consideration when choosing C&S procedures.

Cleaning procedures

Cleaning is one of the most important steps in the cleaning and sanitation process. The cleaning process can be executed as dry cleaning and/or wet cleaning. When done appropriately, cleaning alone can remove over 90 percent of microorganisms.

i) Dry Cleaning

Dry cleaning involves removing visible contaminants and organic materials, such as dirt, manure, bedding, and feed, from working spaces or machinery. It is important to choose the right time for cleaning to avoid releasing viruses into the air. The process includes sweeping, scraping, and using tools like shovels, manure forks, brooms, and brushes to clear surfaces of organic matter and debris. In some cases, large amounts of bedding and manure may require removal and disposal with heavy machinery.

It is advised to clean in the manner described below:

1) Remove as little organic matter as possible; often, the original surface of fans, floors, and walls (such as wood or metal) should still be visible.
2) For advice on how much organic material is appropriate to leave on any surface, connected to whether or not that surface is accessible to restocking animals, consult the unified Incident Command.
3) The third step is to dry or eliminate any remaining moist organic material. This is probably the most recent material to be deposited on the site, making it the most likely to contain infectious material.
4) Make sure the last of the organic material dries; if the weather has rendered the material moist or wet, remove it or let it dry naturally. The virus can be eliminated or destroyed by drying.

5) Dispose of materials properly, adhering to directives from the unified incident command.

6) State and APHIS officials will decide on disposal with the IMT in cases where there is stored manure or manure pits.

ii) Wet Cleaning

Wet cleaning involves the use of water and soap or detergent.

Washing

After removing severe contamination (dry cleaning), surfaces or objects need to be cleansed with detergent. The washing procedure aids in further reducing the amount of bacteria and gets rid of any grease, oil, or exudates that can prevent disinfection from working. One of the C&S process's most usually missed procedures is washing before disinfection.

It is advised to wash in the manners described below:

1) If all electrical power will be switched off for washing, get an alternative power supply.

2) Disconnect any electrical equipment, remove it from the area, and cover it securely with plastic sheeting. In case an electrician is required.

3) In accordance with the site-specific strategy, use brushes to scrub any contaminated surfaces with water and detergent as necessary, making sure the cleansed areas are free of dirt and debris. Organic waste may be removed with the use of warm water. Caked-on substances could need a lot of soaking time. Use water that is warm to hot (90–130°F [32–54°C] or greater).

4) Add up to 40% propylene glycol to water when preparing solutions if surfaces and the surrounding air are below freezing. You may also use heat blankets to keep liquid containers warm.

5) Clean, drain, and flush every part of the feeding and watering systems. Remove and dismantle these gadgets if you can to get rid of organic matter and make room for effective cleaning. Drain, clean, and flush reservoirs.

6) Prior to disinfection, thoroughly clean the fans, casings, motors, belts, curtains, ventilation pads, and louvers on ventilation components to ensure they are free of manure, debris, dust, and filth. Equipment including thermostats, scales, timepieces, electrical panels, switches, and lightbulbs should all be individually cleaned,

sanitized, and protected as necessary against contamination throughout the cleaning process.

7) Use the site-specific disposal strategy to dispose of all C&S solutions.

- **Cleaning agents**

Cleaning agents are substances or products used to remove dirt, dust, stains, and germs from various surfaces. They can be classified into several types based on their composition, purpose, and target surface. Here are some common types of cleaning agents used for cleaning and sanitation:

1) **Water:** Water is a universal solvent and is the basis for most cleaning solutions. It is used as a diluent or carrier for other cleaning agents and is effective in removing certain types of dirt and debris.

2) **Detergents:** Detergents are the most commonly used cleaning agents. They contain surfactants that help break down and remove dirt, grease, and oil. Detergents are available in various forms, including liquid, powder, and gel, and are suitable for cleaning different surfaces such as dishes, floors, and fabrics.

3) **Disinfectants:** Disinfectants are chemicals that kill or deactivate germs, bacteria, viruses, and fungi. They are used to sanitize surfaces and reduce the risk of infections. Common disinfectants include bleach, hydrogen peroxide, quaternary ammonium compounds, and alcohol-based solutions.

4) **Degreasers:** Degreasers are specifically designed to remove grease, oil, and grime from surfaces. They are commonly used in kitchens, garages, and industrial settings. Degreasers may come in the form of sprays, foams, or concentrated liquids.

5) **Abrasives:** Abrasives are substances that help mechanically remove stubborn stains, rust, or deposits from surfaces. They are often used for scrubbing or scouring hard surfaces such as metal, ceramic, or concrete. Examples of abrasives include scouring pads, steel wool, and abrasive powders.

6) **Solvents:** Solvents are chemicals used to dissolve or disperse substances that are not soluble in water. They are effective in removing certain types of stains, adhesives, paints, or markers. Common solvents include acetone, isopropyl alcohol, and mineral spirits.

7) **Glass cleaners:** Glass cleaners are formulated to clean and remove

streaks from glass surfaces such as windows, mirrors, and tabletops. They often contain ammonia or vinegar as the active ingredient.

8) **All-purpose cleaners:** All-purpose cleaners are versatile products that can be used on multiple surfaces. They are designed to clean a wide range of materials, including countertops, appliances, tiles, and laminates. All-purpose cleaners are typically mild and suitable for everyday cleaning.

9) **Acidic cleaners:** Acidic cleaners are used to remove mineral deposits, rust, and scale from surfaces. They are effective on surfaces like toilets, showerheads, and tiles. Common acidic cleaners include vinegar, citric acid, and phosphoric acid-based products.

10) **Enzymatic cleaners:** Enzymatic cleaners contain specific enzymes that break down organic matter such as stains, blood, urine, or food residues. They are often used in healthcare facilities, veterinary clinics, or for treating stains on fabrics.

- **Disinfection technique**

Disinfection techniques are employed to eliminate or inactivate microorganisms and reduce the risk of infections. Several common disinfection techniques used for cleaning and sanitation include:

1) **Surface wiping:** This technique involves using disinfectant wipes or cloths to wipe down surfaces. The disinfectant solution is applied to the wipe, and the surface is thoroughly wiped to ensure proper coverage. It is effective for disinfecting frequently touched surfaces like doorknobs, countertops, and electronic devices.

2) **Spray and wipe:** In this technique, a disinfectant solution is sprayed directly onto the surface and then wiped off with a cloth or paper towel. It is commonly used for larger surfaces like tables, chairs, and equipment.

3) **Immersion:** This technique is used for disinfecting smaller items that can be immersed in a disinfectant solution. The items are placed in a container filled with the disinfectant, ensuring complete submersion. After a specified contact time, the items are removed and allowed to air dry.

4) **Fogging or misting:** Fogging or misting involves using specialized equipment to generate a fine mist or fog of disinfectant solution. The mist is sprayed into the air or directed onto surfaces, allowing it to settle and disinfect the area. It is commonly used for disinfecting

large spaces, such as offices, classrooms, or healthcare facilities.

5) **UV-C disinfection:** Ultraviolet-C (UV-C) light is used to disinfect surfaces and air. UV-C light damages the DNA or RNA of microorganisms, preventing their replication and rendering them inactive. UV-C disinfection systems are often used in healthcare settings, laboratories, and food processing facilities.

6) **Electrostatic spraying:** Electrostatic sprayers are used to apply a charged disinfectant mist to surfaces. The charged particles attract and cling to surfaces, providing even coverage. This method is efficient for disinfecting large areas quickly, such as in public transportation or stadiums.

7) **Steam cleaning:** Steam cleaning utilizes high-temperature steam to kill microorganisms and sanitize surfaces. The steam is applied through a specialized machine that heats water to produce steam, which is then directed onto the surface. It is effective for cleaning and disinfecting surfaces like floors, carpets, and upholstery.

- **Disinfectants for the meat industry**

In the meat industry, proper sanitation and disinfection are crucial to maintain food safety and prevent the spread of harmful microorganisms. Here are some common disinfectants used in the meat industry along with their explanations:

1) **Chlorine compounds:** Chlorine-based disinfectants, such as sodium hypochlorite (bleach) or calcium hypochlorite, are widely used in the meat industry. They are effective against a broad range of microorganisms and can be used for sanitizing surfaces, equipment, and utensils. However, it's important to use them at the appropriate concentration and ensure thorough rinsing to prevent any residual chlorine taste or odor.

2) **Quaternary ammonium compounds (Quats):** Quats, such as benzalkonium chloride or dodecylbenzenesulfonic acid, are commonly used disinfectants in the meat industry. They have good detergent properties and are effective against various bacteria and some viruses. Quats can be used for sanitizing food contact surfaces, floors, and equipment. It's important to follow the manufacturer's instructions for proper dilution and contact time.

3) **Peroxyacetic acid (PAA):** PAA is a powerful disinfectant and sanitizer that is effective against bacteria, fungi, and viruses. It is often used in the meat industry for sanitizing equipment, conveyors,

and processing surfaces. PAA is known for its fast action, relatively low toxicity, and ability to break down into harmless byproducts. However, proper protective measures and dilution guidelines should be followed during its use.

4) **Hydrogen peroxide:** Hydrogen peroxide is a versatile disinfectant that can be used in the meat industry. It has broad-spectrum antimicrobial activity against bacteria, viruses, yeasts, and molds. Hydrogen peroxide is often used for fogging or misting applications to disinfect large areas or processing rooms. It decomposes into water and oxygen, leaving no harmful residues.

5) **Acidified sodium chlorite (ASC):** ASC is a chlorine-based disinfectant that is commonly used in the meat industry for surface sanitizing and controlling microbial growth. It is effective against various bacteria, including pathogens like Salmonella and E. coli. ASC is usually applied as a solution or spray and requires proper contact time and rinsing to ensure efficacy and food safety.

6) **Iodine compounds:** Iodine-based disinfectants, such as iodophors, are used in the meat industry for their antimicrobial properties. They are effective against bacteria, fungi, and viruses. Iodophors are often used for sanitizing hands, surfaces, and equipment. However, their use should be closely monitored as they may cause staining and can be corrosive to certain metals.

It's important to note that the specific choice of disinfectant in the meat industry depends on various factors, including the target microorganisms, surface or equipment being treated, and regulatory requirements.

- **Cleaning and disinfection plans**

Developing a comprehensive cleaning and disinfection plan is essential to maintain a clean and safe environment. Here are some key steps and considerations to include in a cleaning and disinfection plan:

1) **Identify high-touch surfaces:** Identify the surfaces and areas that are frequently touched or come into direct contact with food, such as countertops, cutting boards, utensils, equipment handles, and switches. These surfaces should be prioritized for regular cleaning and disinfection.

2) **Select appropriate cleaning agents and disinfectants:** Choose cleaning agents and disinfectants that are suitable for the specific surfaces and areas being cleaned. Consider the type of contaminants,

target microorganisms, and any regulatory requirements. Follow the manufacturer's instructions for proper use, dilution ratios, and contact times.

3) **Establish cleaning frequencies:** Determine the appropriate frequency for cleaning and disinfection based on factors such as the level of food handling, foot traffic, and the risk of contamination. High-traffic areas or surfaces may require more frequent cleaning compared to low-traffic areas.

4) **Develop standard operating procedures (SOPs):** Create clear and detailed SOPs for cleaning and disinfection activities. Include step-by-step instructions for each task, such as pre-cleaning, application of cleaning agents/disinfectants, contact time, rinsing, and drying. Train staff on these procedures to ensure consistency and adherence.

5) **Provide adequate training:** Train employees on proper cleaning and disinfection techniques, including the correct use of cleaning agents, dilution ratios, safety precautions, and the importance of following SOPs. Regularly reinforce training to maintain good hygiene practices.

6) **Implement personal protective equipment (PPE):** Identify the necessary PPE for cleaning and disinfection tasks, such as gloves, goggles, aprons, or masks. Ensure that employees understand when and how to use PPE correctly to protect themselves and prevent cross-contamination.

7) **Establish cleaning checklists:** Develop checklists to ensure that all necessary cleaning tasks are completed. This can include a list of areas, surfaces, and equipment to be cleaned, along with checkboxes or timestamps to track completion. Regular inspections can help monitor compliance and identify any areas needing improvement.

8) **Document and review:** Keep detailed records of cleaning and disinfection activities, including dates, times, products used, and any issues encountered. Regularly review these records to assess the effectiveness of the cleaning plan, identify trends, and make any necessary adjustments.

9) **Monitor and audit:** Implement a monitoring and auditing system to evaluate the effectiveness of the cleaning and disinfection plan. Conduct periodic inspections, microbial testing if required, and gather feedback from staff to ensure compliance and identify areas for improvement.

10) **Stay informed:** Stay updated on relevant regulations, guidelines, and best practices related to cleaning and disinfection in the specific industry. Stay informed about new products or techniques that may improve efficiency or enhance safety.

Conclusion

In conclusion the importance of cleaning and sanitation in various settings for maintaining hygiene, preventing the spread of diseases, and ensuring safety. It emphasizes the need for effective cleaning practices, including the removal of dirt and contaminants from surfaces, followed by the use of disinfectants to eliminate microorganisms. A well-designed cleaning and sanitation plan is crucial, involving identifying high-touch surfaces, selecting appropriate cleaning agents, establishing frequencies, providing training, and following industry guidelines. Consistent adherence to these protocols significantly reduces the risk of infections and cross-contamination. Regular review and updates based on new information and technologies are essential. The commitment to thorough cleaning, proper disinfection, and a culture of cleanliness promotes a safer and healthier environment for all.

References

1. Missouri Department of Agriculture. (2008). Cleaning and Disinfection: Standard Operating Guide No. 004. In Missouri Department of Agricultural Emergency Response Actions: Livestock Disease Emergency.

2. U.S. Department of Labor Occupational Safety and Health Administration. (1998). In Appendix B-2 to § 1910.134: Respirator Cleaning Procedures.

3. USDA APHIS and Center for Food Security and Public Health (CFSPH). (2014). FAD PReP/NAHEMS Guidelines: C&D. Retrieved from APHIS FAD PReP.

4. Vaid NR and Verma R. (2022). Change in lifestyle and dietary habits during COVID-19 pandemic. In*t. J. of Adv. Res.* 10: 576-581.

5. Abu Al-Rub, Fahmi & Fahmi, Abu & Al-Rub, & Shibhab, Penelope & Al-Rub, Safwan & Paparella, Antonello. (2020). Cleaning and Disinfection. 10.29011/978-1-951814-05-2-006.

Chapter - 14
Meat Processing Techniques

Authors

Akash Khorde
MTech., MIT ADT, Pune, Maharashtra, India

Dhaneshwari Karnavat
MTech., MIT ADT, Pune, Maharashtra, India

Deepti Chaudhari
Assistant Professor, MIT ADT, Pune, Maharashtra, India

Chapter - 14

Meat Processing Techniques

Akash Khorde, Dhaneshwari Karnavat and Deepti Chaudhari

Abstract

Poultry meat processing is a critical aspect of ensuring the safety, quality, and overall satisfaction of poultry products. This chapter provides a comprehensive overview of key elements pertaining to poultry meat processing, encompassing techniques, carcass preparation, handling and storage of frozen poultry, thawing methods, and the significance of proper cooking practices. The chapter briefly discusses the historical development of poultry meat processing and underscores its pivotal role in meeting consumer expectations. It emphasizes the paramount importance of poultry meat processing in guaranteeing food safety, mitigating the risk of foodborne illnesses and enhancing the gustatory experience associated with poultry products. The various techniques employed in poultry meat processing including refrigeration, freezing, smoking, curing, dehydration, freeze drying, and canning are explored in detail. Each technique is expounded upon, delineating their merits, drawbacks and their indispensable role in preserving and preparing poultry products.

1. Introduction

1.1 Brief history of poultry meat processing

In the past 50 years, the poultry industry has seen significant market growth and changes. While the demand for oven-ready carcasses was once high in developed countries has now been replaced by a preference for a wider range of further-processed convenience items. As a result, poultry companies must focus on food technology, product development and meat production. Consumers have also become more knowledgeable about issues such as food safety, animal welfare, and environmental protection, leading to increased expectations and the implementation of new legislation and production schemes in Europe. These developments have placed a significant demand on the industry requiring a strong knowledge base and appropriate research to support all aspects of its activities (Mead, G. (Ed.). 2004).

Over the past three decades, the global chicken meat industry has undergone rapid expansion fuelled by modernization and automation as well as genetic advancements that enable greater yield. While the consumption of various types of meat has increased in absolute and per capita terms, the production of chicken meat has been the most successful in terms of standardizing its processes, boosting yields, and achieving significant cost and price reductions. This success is expected to continue with forecasts indicating further expansion of chicken meat production both locally and internationally. (OECD 2018).

Poultry farming has emerged as one of the fastest-growing segments of the agricultural sector in India, with an annual growth rate of approximately eight percent. Over the past four decades, the poultry sector has significantly transformed from a backyard activity to a major commercial agri-based industry. This transformation has been made possible by the constant efforts in upgrading, modifying, and applying new technologies in the sector. As a result, the growth in poultry and allied sectors has been many fold and multifaceted. The industry's adoption of modern technologies has enabled it to achieve significant improvements in productivity, efficiency and profitability. (Chatterjee, R. N., & Rajkumar, U. 2015).

As income levels increase, so does the demand for meat. This trend is evident in the consumption of poultry meat across the world. In 2011, the average consumption of poultry meat per capita was 14.5 kg, which represented a 32% increase from the year 2000, when the average consumption was 11 kg. In Asia, the consumption of poultry meat per capita per year increased from 6.6 kg to 9.5 kg during the same period. By 2015, meat consumption worldwide had reached 41.3 kg per capita, and it is projected to increase to 45.3 kg per capita by 2030. Poultry meat is expected to contribute 17.2 kg or 38% to total meat consumption in 2030, compared to 13.8 kg or 33% in 2015. This indicates a significant increase in the contribution of poultry meat to total meat consumption, from 33% to approximately 38%. Developing countries are also expected to see a rise in poultry meat consumption, from 10.5 kg per capita in 2015 to 14 kg per capita by 2030, which represents a 25% increase. (Wahyono, N. D., & Utami, M. M. D. 2018).

2. Importance of poultry meat processing

Poultry meat, particularly chicken meat, is considered to be white meat and is known to contain highly valuable proteins and amino acids essential for human health. On the other hand red meat refers to the meat of other domesticated animals and is classified separately due to differences in its nutritional value. Many studies conducted across the world on varying food consumption habits have consistently demonstrated the link between human nutrition and overall health. Such studies have also highlighted the importance of including nutrient-rich foods like chicken in our diets to maintain optimal health (Jilo and Hasan 2022).

The poultry industry has experienced significant growth in recent years, especially in the development of value-added and further-processed products. In the United States, for instance, only a small fraction, less than 10%, of all broilers are sold as whole carcasses. This shift towards further-processed products has been the most crucial area of growth in the poultry industry. (Mead, G. (Ed.). 2004).

The demand for primary poultry products including meat and eggs has been steadily increasing across the world due to consumer preference for high-quality products at relatively low prices. This demand has led to a rise in production which has been efficiently managed to meet the growing needs of consumers. Between the years 1995 to 2005, there was a significant increase in the consumption and production of poultry products worldwide. Chicken meat, in particular, witnessed a remarkable 53% increase in production and consumption while turkey meat consumption increased by 13%. Similarly, the consumption of duck meat and goose meat rose by 67% and 53%, respectively during this period. The production and consumption of chicken eggs also rose by 39% while the consumption of other types of eggs increased by 27%. These trends highlight the importance of poultry production and its ability to meet the growing demand for high-quality protein sources. The information is crucial for anyone interested in understanding global food trends and the role of poultry products in meeting the needs of consumers.

2. Poultry meat processing techniques

2.1 Refrigeration

2.1.1 Temperature and storage condition

Refrigeration is considered the most popular and effective method for the short-term preservation of meat. This technique involves cooling the meat to temperatures below the optimal range for microbial growth, enzymatic reactions, and chemical reactions, thereby slowing or limiting the rate of spoilage. The storage of fresh meat is typically carried out at a refrigeration temperature range of 2 to 5°C. Chilling is crucial for maintaining meat hygiene, safety, shelf life, appearance, and nutritional quality. By using refrigeration, meat can be stored for a longer period of time while still maintaining its quality and freshness. This information is essential for individuals who want to learn about meat preservation techniques and for professionals in the food industry who are responsible for maintaining food safety and quality. (Pal *et al.*, 2017)

A temperature range of 5-60°C is known as the "danger zone" for meat storage due to the rapid growth of bacteria within this range. The optimal temperature for bacterial growth in meat is between 25-45°C. This information emphasizes the critical importance of proper temperature control in meat storage to prevent the growth of harmful bacteria that can cause foodborne illnesses (Jay *et al.*, 2008).

Rapid cooling of meat to a temperature of 0-2°C within 2-4 hours after slaughter can improve its quality and extend its shelf life. This emphasizes the importance of proper refrigeration techniques in meat processing to ensure high-quality and safe products for consumers. (Schirmer and Lorenzen 2019)

2.1.2 Importance of proper refrigeration

Enzymatic reactions, which can cause discoloration and off-flavors in meat, can also be slowed down by refrigeration. For example, the oxidation of fat in meat can cause rancidity and spoilage, but refrigeration can slow down this process and help preserve the freshness of the meat (Kerry & Kerry, 2002).

It was found that refrigerated storage at a temperature of 4°C (39.2°F) can significantly reduce the growth of harmful bacteria in ground beef. The study concluded that proper refrigeration is essential for maintaining the safety of ground beef, emphasizing the importance of maintaining proper refrigeration practices in the food industry to prevent the growth of bacteria and ensure the safety of consumers (Rhee *et al.*, 2013).

2.2 Freezing

Freezing is a popular technique employed for food preservation which entails bringing the temperature of the food down to a point below freezing. This method helps to impede the growth of microorganisms and enzymes responsible for food spoilage (Bansal, Chhillar, & Mangal, 2014). Freezing also serves to retain the texture, color, and nutritional value of food products such as fruits and vegetables (Beck, 2017).

Freezing can be accomplished by utilizing various methods such as air-blast freezing, plate freezing and cryogenic freezing (Bansal *et al.*, 2014). Air-blast freezing is a frequently employed technique utilizes cold air that is blown over food to rapidly cool it (Fellows, 2009). Plate freezing on the other hand, entails placing food on a flat surface that is chilled by the refrigerant (Fellows, 2009). Cryogenic freezing, a method that utilizes very cold gases such as liquid nitrogen or carbon dioxide is employed to swiftly freeze food (Bansal *et al.*, 2014).

2.3 Advantages and disadvantages of freezing

2.3.1 Advantages

1) **Longer shelf life:** Freezing can extend the shelf life of food by preserving its quality for a longer period of time (Singh, 2014).
2) **Nutritional value:** Freezing helps to maintain the nutritional value of food as it prevents the loss of vitamins and minerals that can occur during other methods of preservation (Beck, 2017).
3) **Convenience:** Freezing food can provide a convenient and time-saving way to prepare meals as it allows for bulk cooking and easy storage (Singh, 2014).

2.3.2 Disadvantages

1) Freezing has some drawbacks, which include changes in texture and flavor in certain foods due to the expansion of water when it freezes, leading to cell damage (Beck, 2017).
2) Furthermore, over time, ice crystals can impair the structure of the food and affect its texture and flavor, leading to a decrease in food quality (Singh, 2014).
3) Another drawback of freezing food is its significant energy consumption. It necessitates the use of electricity to power freezers and the transportation of frozen goods, resulting in additional energy consumption (Beck, 2017).

2.4 Smoking

2.4.1 Types of smoking

There are three types of smoking as follows.

- **Cold smoking:** This smoking method involves smoking food for extended periods at temperatures between 68°F and 86°F (20°C to 30°C) sometimes up to 24 hours. Cold smoking is commonly used for foods such as cheese, fish, and meat (Sánchez-Escalante, Torrescano, & Mota-Rojas, 2011).
- **Hot smoking:** This smoking method involves smoking food at higher temperatures ranging from 126°F to 176°F (52°C to 80°C) for

shorter periods usually a few hours. Hot smoking is commonly used for foods such as beef, pork, and poultry (Fellows, 2009).

- **Liquid smoking:** This smoking method involves immersing food in a solution containing liquid smoke or spraying liquid smoke onto the food. It is commonly used for processed meat products like bacon and sausages (Lafarga & Hayes, 2017).

2.4.2 Importance of smoking

Smoking is a traditional method of food preservation and flavor enhancement that has been used for centuries (Sánchez-Escalante, Torrescano, & Mota-Rojas, 2011). Smoking can preserve food by reducing water activity and creating an environment that inhibits the growth of bacteria and other microorganisms. Smoke also imparts a unique flavor to food that is highly valued by consumers and the complex flavors of smoke can come from a variety of sources, such as the type of wood used, the temperature of the smoke and the length of time that the food is smoked (Lafarga & Hayes, 2017).

Smoking is a versatile technique that can be used for a wide variety of food products, including meat, fish, cheese, and vegetables. It is an important part of traditional barbecue and meat-curing practices in the United States and is also used in the production of specialty foods like smoked salmon in Europe (Fellows, 2009).

Apart from its culinary significance, smoking also has cultural and traditional importance in many parts of the world. In some cultures, smoking is an important part of religious ceremonies and celebrations (Lafarga & Hayes, 2017). In addition, smoking has been shown to have a positive effect on the mental well-being of individuals as it can induce feelings of relaxation and stress relief (Nakajima *et al.,* 2019).

2.5 Curing

2.5.1 Types of curing

1) **Dry curing:** Salt and other seasonings are applied to the surface of the meat which is then allowed to cure in a cool, dry environment for several weeks or months. This method is frequently used for cured meats such as bacon and ham (Ruiz-Carrascal, 2010).

2) **Wet curing:** In wet curing, meat is submerged in a solution of brine containing salt, nitrites, and other flavorings for a few days to several weeks. This method is typically used for cured products such as corned beef and pastrami (Ruiz-Carrascal, 2010).

3) **Combination curing:** Combination curing involves using both dry and wet curing methods. The meat is coated with a dry mixture and then placed in a brine solution. This approach is frequently employed for bacon and ham production (Ruiz-Carrascal, 2010).

4) **Injection curing:** The curing mixture is injected into the meat using a needle or another tool in this method. This allows for a faster curing process and is suitable for larger cuts of meat (Ruiz-Carrascal, 2010).

5) **Sugar curing:** Sugar is added to the curing mixture to balance the saltiness and provide a sweet flavor to the meat in sugar curing. This approach is commonly used in the production of hams (Ruiz-Carrascal, 2010).

2.5.2 Importance of curing

Curing is a traditional method of food preservation that has been utilized for centuries to prevent the growth of harmful microorganisms and increase the shelf life of food products. The primary objective of curing is to protect food from bacterial spoilage and maintain food safety (Sharma *et al.,* 2016). Moreover, curing also contributes to the sensory attributes of meat products by adding distinctive flavors and textures. The addition of salt and other seasonings during the curing process enhances the taste of meat, making it more appetizing to consumers (Sharma *et al.,* 2016).

Curing is also significant in the production of various traditional and specialty meat products. For example, cured meats such as ham, bacon, and corned beef have been widely consumed throughout history and are still popular today. Moreover, curing is employed in the creation of exceptional meat products like prosciutto and salami in Europe (Ruiz-Carrascal, 2010).

2.6 Dehydration

2.6.1 Oven dehydration

Oven dehydration is a widely employed method of removing moisture from various materials, such as fruits, vegetables, and herbs. It involves placing the material inside an oven and subjecting it to low heat, which facilitates the evaporation of moisture. This technique is commonly utilized in food preservation to extend the shelf life of perishable items by significantly reducing their water content.

2.6.2 Vacuum dehydration

Vacuum dehydration is a method used to eliminate moisture from heat-sensitive materials by placing them in a vacuum chamber and reducing the pressure to lower the boiling point of water. As a result, water evaporates at a lower temperature than it would under normal atmospheric conditions. This technique is commonly employed for the removal of moisture from delicate substances that may be damaged or undergo undesirable changes when exposed to high temperatures.

2.7 Freeze drying

2.7.1 Process of freeze drying

Freeze drying also known as lyophilization is a widely used dehydration process in industries such as pharmaceuticals, food preservation and biotechnology. It involves three key stages: freezing, primary drying and secondary drying. During the freezing stage, the substance is cooled below its freezing point causing water within it to form ice crystals. Controlled freezing rates and conditions are essential to maintain the substance's structural integrity. In the primary drying stage, the pressure is reduced and gentle heat is applied causing the ice crystals to undergo sublimation and transforming directly from a solid to a vapor state without becoming liquid. This process removes the water molecules as vapor leaving behind a porous structure. In the final stage of secondary drying any remaining moisture trapped in the material or adsorbed onto its surface is removed by slightly raising the temperature still below the substance's melting point. The duration of the secondary drying stage depends on the material's specific requirements and desired moisture content.

Freeze drying offers several advantages over other dehydration methods. Firstly, it enables the long-term preservation of sensitive materials by minimizing chemical and physical degradation that can occur with alternative drying methods. Secondly, the low temperature used in freeze drying helps retain the substance's original properties, including taste, aroma, and biological activity, which is vital in applications such as pharmaceuticals and food preservation. Lastly, freeze-dried products possess a lightweight and porous structure allowing them to rapidly rehydrate when exposed to water. This feature is beneficial in situations where quick reconstitution is necessary, such as in the medical field or instant food preparation. Overall, freeze-drying is a versatile technique for preserving and stabilizing sensitive materials while maintaining their quality and functionality.

2.7.1 Importance of freeze drying

Freeze drying also known as lyophilization is a critical process with diverse applications across several industries. Its significance lies in its capacity to preserve and stabilize delicate materials while maintaining their structural integrity and biological activity. In the pharmaceutical industry, freeze-drying is particularly vital as it enhances the stability and shelf life of medications and vaccines. By removing water through freeze drying, pharmaceutical products can be stored for extended periods without degradation, ensuring their effectiveness and potency. Additionally, freeze drying enables the production of easily reconstitute drugs, making them convenient for parenteral administration.

In the food industry, freeze-drying plays a crucial role in preserving perishable food items while retaining their nutritional value, flavor, and texture. This process removes water from the food, resulting in freeze-dried products with an extended shelf life. These lightweight and compact food items are convenient for transportation and storage. Moreover, the rehydration properties of freeze-dried foods make them ideal for quick and easy preparation in emergency situations, outdoor activities, and even space travel. The ability of freeze drying to preserve the quality and functionality of sensitive biomolecules also makes it an invaluable technique in biotechnological advancements, allowing for the long-term storage and subsequent use of enzymes, proteins, and other biological materials in various research, diagnostic, and therapeutic applications.

2.8 Canning

2.8.1 Process of canning

Canning is a food preservation technique that involves sealing food in airtight containers such as cans or jars, and subjecting them to heat to eliminate microorganisms, enzymes and other spoilage-causing agents. By creating a sealed environment, canning extends the shelf life of food by inhibiting the growth of bacteria, yeast and molds. This process allows canned foods to be stored for long periods without refrigeration while preserving their nutritional value, flavor, and texture. The canning process typically includes steps such as food preparation, heat treatment to eliminate harmful microorganisms, packing the food into sterilized containers, and sealing them with airtight lids. Subsequently, the sealed containers undergo additional heating, commonly known as retorting, to eliminate any remaining microorganisms and enzymes. The history of canning dates back to the early 19th century when Nicolas Appert, a French confectioner, pioneered this preservation method and was awarded a prize in 1809 for his successful canning experiments. Since then, ongoing advancements in canning technology and equipment have enhanced the efficiency and safety of the process, revolutionizing the food industry.

Food Preparation (Cleaning, peeling, cutting)

↓

Heat Treatment (Eliminating microorganisms)

↓

Packing Food into Sterilized Containers

↓

Applying Sealing Material (e.g., Metal Lid)

↓

Sealing the Containers

↓

Retorting

↓

Labelling and Packaging

2.8.2 Advantages and disadvantages of canning

2.8.2.1 Advantages of canning

1) Extended Shelf Life: Canning enables long-term preservation of food, extending its shelf life. When properly canned, food can last for a year or more depending on the canning method and the type of food. This is particularly beneficial for seasonal produce or situations where fresh food availability is limited (USDA, 2020).

2) Nutrient Retention: The high heat involved in canning destroys harmful bacteria and enzymes that can lead to spoilage. This heat treatment helps preserve essential vitamins and minerals, retaining the nutritional value of the food being canned (Ball, 2014).

3) Convenience: Canned foods offer convenience in storage and usage. Once properly canned, they do not require refrigeration freeing up space in the refrigerator. Canned foods are also portable, making them convenient for outdoor activities like camping, hiking, and emergency situations (USDA, 2020).

4) Cost-Effectiveness: Canning can be a cost-effective method of preserving food, particularly when dealing with bulk or garden-fresh produce. It allows individuals to take advantage of seasonal abundance, reducing food waste and saving money (Ball, 2014).

5) Food Safety: When carried out correctly, canning provides a safe and

reliable means of preserving food. The heat treatment involved in the canning process eliminates harmful bacteria and microorganisms, reducing the risk of foodborne illnesses (USDA, 2020).

2.8.2.2 Disadvantages of canning

1) **Initial investment:** Canning involves an initial investment in equipment such as canners, jars, lids and utensils. While these tools can be reused for multiple canning sessions, the upfront cost may discourage some individuals from adopting this preservation method (Ball, 2014).

2) **Time-consuming:** Canning can be a time-consuming process, especially when compared to other food preservation methods. It includes various steps like preparation, processing, and cooling. Additionally, maintaining proper hygiene and following recommended procedures add to the time required for canning.

3) **Quality and texture changes:** Certain foods may undergo texture and quality changes during the canning process. For example fruits may become softer and vegetables may lose some of their crispness. The high heat used in canning can also lead to flavor changes or nutrient loss, although these effects are generally minimal if the process is conducted correctly (Ball, 2014).

4) **Risk of spoilage:** Despite the advantages, there is a risk of spoilage if the canning process is not performed properly. Insufficient sealing or improper processing can result in microbial contamination, leading to food spoilage or even severe food poisoning such as botulism. It is crucial to follow recommended canning guidelines and use tested recipes to minimize these risks.

5) **Storage space:** Canned foods require storage space as they are bulkier compared to fresh or frozen foods. Depending on the quantity of food being canned, adequate space needs to be allocated in the pantry or storage area.

3. Preparation of carcasses for cooking

3.1 Introduction

Cleaning and washing are crucial processes within meat processing facilities, playing a vital role in ensuring food safety, preventing cross-contamination and maintaining high hygiene standards.

3.1.1 Food safety

1) Food Safety: Proper cleaning and washing protocols are essential for eliminating potential hazards such as pathogenic bacteria (e.g., *Salmonella, Escherichia coli*) and foodborne pathogens.

2) Cross-Contamination Prevention: Thorough cleaning and washing procedures are crucial in preventing cross-contamination between different stages of meat processing, equipment, and surfaces.

3) Regulatory Compliance: Meat processing facilities are subject to strict sanitation requirements set by regulatory bodies such as the United States Department of Agriculture (USDA) and the Food and Drug Administration (FDA).

3.2 Trimming

3.2.1 Introduction

Trimming plays a crucial role in meat processing as it entails the removal of unwanted components like excess fat, connective tissue, and undesirable portions from meat cuts. This essential step not only enhances the quality and visual appeal of meat products but also contributes to improved flavour, tenderness and overall consumer satisfaction. By effectively trimming meat cuts, processors can achieve a desirable balance of fat content, improve taste, ensure tenderness, and meet consumer expectations for high-quality meat.

3.2.2 Product quality and appearance

1) Product Quality and Appearance: This meticulous trimming process enhances the overall appearance and aesthetic appeal of the meat, contributing to a visually appealing product that meets consumer expectations for quality.

2) Flavour Enhancement: By removing excess fat, the likelihood of greasiness is minimized, enabling the natural flavours of the meat to shine through.

3) Tenderization: Trimming tough connective tissues significantly enhances the tenderness of meat cuts.

4) Consistency and Yield: Maintaining consistent trimming practices ensures that portion sizes and product specifications remain uniform.

4. Frozen poultry

4.1 Advantages and disadvantages of frozen poultry

4.1.1 Advantages of Frozen Poultry

1) Convenience and Availability: Frozen poultry is easily accessible in

grocery stores year-round, providing consumers with the convenience of purchasing and storing it for future use. It simplifies meal planning and reduces the need for frequent shopping trips.

2) Extended Shelf Life: Freezing poultry preserves its quality and nutritional value over an extended period. It can be stored for months without significant changes in taste, texture, or overall quality, which helps in minimizing food waste.

3) Safety: Freezing poultry at sufficiently low temperatures inhibits the growth of bacteria and other microorganisms that can cause foodborne illnesses. It ensures the safety of the product until it is consumed.

4) Cost-Effectiveness: Frozen poultry is generally more affordable compared to fresh poultry. It enables consumers to save money by purchasing in bulk or taking advantage of promotional offers.

4.1.2 Disadvantages of Frozen Poultry

1) Texture and Flavour Changes: The freezing and thawing process can sometimes result in alterations in the texture and flavour of poultry. Ice crystal formation during freezing may affect the texture, and there may be a slight loss of moisture during thawing.

2) Nutritional Loss: Although freezing can preserve the nutritional value of poultry to a large extent, some vitamins and nutrients may be lost during the freezing process. However, the overall impact on nutritional content is typically minimal.

3) Thawing Time: Properly thawing frozen poultry requires planning and time. Depending on the size, it may take several hours or overnight to thaw in the refrigerator. Improper thawing techniques such as leaving it at room temperature can increase the risk of bacterial growth.

4) Quality Variations: The quality of frozen poultry can vary based on factors like storage conditions, freezing methods and packaging. Poorly packaged or stored poultry may experience freezer burn which can impact its taste and texture.

4.2 Handling and storage of frozen poultry

Packaging: To maintain the quality and prevent freezer burn, proper packaging of frozen poultry is essential. It is recommended to use moisture-proof and airtight materials like freezer bags or containers which minimize exposure to air and moisture.

Freezing Guidelines: Follow these guidelines when freezing poultry:

Ensure the poultry is fresh and free from signs of spoilage before freezing. Divide larger portions into smaller, freezer-friendly portions for faster freezing and easier thawing.

Label the packages with the date of freezing to track storage time accurately.

Storage Temperature: Frozen poultry should be stored at a constant temperature of -18°C (0°F) or below to ensure its quality and safety.

Storage Duration: The recommended storage duration for frozen poultry depends on the type of poultry and the quality of freezing. Whole poultry can typically be stored for up to 12 months, while poultry parts and ground poultry can be stored for 3 to 4 months. However, it's important to consult specific packaging or guidelines for precise storage durations.

Thawing: Proper thawing is vital to maintain the quality and safety of frozen poultry. The recommended method is to thaw the poultry gradually in the refrigerator. Allow sufficient time, typically around 24 hours for every 2.3 to 2.7 kg (5 to 6 pounds) of poultry, for thorough thawing.

Refreezing: While it's generally safe to refreeze poultry that has been thawed in the refrigerator but not cooked, repeated cycles of thawing and refreezing can affect the quality, resulting in potential texture and flavour changes. It's best to avoid refreezing whenever possible.

5. Thawing of poultry

5.1 Methods of thawing

1) Refrigerator Thawing: The refrigerator thawing method involves placing the frozen food item in the refrigerator to thaw slowly over time. This method is considered one of the safest as it maintains a consistent and controlled temperature, reducing the risk of bacterial growth. However, it is also the slowest method and may take several hours or even days, depending on the size and thickness of the food item.

2) Cold Water Thawing: Cold water thawing entails sealing the frozen food item in a leak-proof plastic bag and submerging it in cold water. To maintain a safe temperature, the water should be changed every 30 minutes. While this method is faster than refrigerator thawing, it requires diligent attention and monitoring to ensure that the water remains cold (below 40°F/4°C) and that the food item does not enter the "danger zone" where bacterial growth can occur.

3) Microwave Thawing: Microwave thawing is a quick method commonly used for small food items. It should be approached with caution and close supervision. Most microwaves have a defrost setting or power level specifically designed for thawing. Following the microwave manufacturer's instructions is essential and it is recommended to rotate or flip the food item frequently to ensure even thawing.

4) Cooking from Frozen: Certain food items, such as select vegetables or pre-packaged meals, can be cooked directly from a frozen state without prior thawing. This method is convenient but may require longer cooking times to ensure thorough cooking.

5.2 Importance of proper thawing

Proper thawing is essential for preventing the growth of harmful bacteria and minimizing the risk of foodborne illnesses.

Safe thawing practices help limit the time food spends in the temperature danger zone (40°F/4°C to 140°F/60°C), where bacteria can multiply rapidly.

Refrigerator thawing and cold water thawing are recommended methods as they enable a controlled and gradual temperature increase, which helps reduce bacterial growth.

Slow and controlled thawing helps preserve the original quality, flavour and texture of the food product.

Rapid thawing methods like microwaving can lead to uneven heating, partial cooking, and a loss of texture in certain food items.

6. Cooking of poultry

6.1 Methods of cooking

1) Roasting: Poultry is cooked in an oven at a higher temperature, resulting in a crispy exterior and tender meat. It is important to cook poultry to a minimum internal temperature of 165°F (74°C) to ensure food safety.

2) Grilling: Poultry is cooked over direct heat, imparting a smoky flavor and producing crispy skin. Thorough cooking to an internal temperature of 165°F (74°C) is essential for food safety.

3) Frying: Poultry is immersed in hot oil, creating a crispy and golden-brown exterior. Monitoring the oil temperature and ensuring poultry reaches an internal temperature of 165°F (74°C) is important.

4) Baking: Poultry is cooked in an oven, often with marinades, sauces, or stuffing. Following specified cooking times and temperatures is crucial for thorough cooking and food safety.

5) Braising: Poultry is cooked in a liquid, like broth or wine, at a low temperature for an extended period, resulting in tender and flavorful meat. Poultry should reach an internal temperature of 165°F (74°C) for safety.

6) Stir-Frying: Small, thinly sliced pieces of poultry are quickly cooked in a hot pan with oil and vegetables, retaining moisture and flavor. Thoroughly cooking poultry to an internal temperature of 165°F (74°C) is necessary.

6.2 Importance of proper cooking

Elimination of Pathogens: Cooking poultry to the appropriate internal temperature helps eliminate harmful bacteria, such as Salmonella and Campylobacter, commonly present in raw poultry. These bacteria can cause severe illnesses, including gastroenteritis, fever, abdominal cramps, and potentially serious complications

Prevention of Foodborne Illnesses: Consuming undercooked or raw poultry increases the risk of foodborne illnesses. Symptoms of these illnesses can range from mild discomfort to severe complications, particularly among vulnerable populations such as children, the elderly, and individuals with weakened immune systems

Destruction of Parasites: Poultry may harbor parasites like Toxoplasma and Trichinella, which can lead to infections when poultry is consumed undercooked or raw. Adequate cooking destroys these parasites, ensuring that poultry is safe to eat.

Texture and Flavour: Proper cooking techniques contribute to desirable texture and flavour in poultry. Cooking poultry to the correct internal temperature ensures even cooking, resulting in tender, juicy, and enjoyable poultry.

7. Conclusion

In conclusion, the application of poultry meat processing techniques, encompassing refrigeration, freezing, smoking, curing, dehydration, freeze drying, and canning, is of paramount importance in the preservation and preparation of poultry products. Adherence to proper handling practices, meticulous temperature control, and diligent execution of thawing and cooking methods are pivotal in ensuring food safety, upholding product quality, and elevating the sensory attributes of poultry. By diligently following recommended guidelines and comprehending the significance of each procedural aspect, we can secure the gratification, and optimal culinary experience associated with poultry meat, while concurrently adhering to industry benchmarks and standards.

References

1. Ball, J. (2014). Ball Complete Book of Home Preserving. Robert Rose Incorporated.
2. Bansal, N., Chhillar, S., & Mangal, M. (2014). Methods of food preservation. In Food preservation and safety (pp. 45-72). Springer, New Delhi.
3. Beck, L. (2017). Freezing foods. Colorado State University Extension. Retrieved from https://extension.colostate.edu/topic-areas/nutrition-food-safety-health/freezing-foods-9-346/
4. Beck, L. (2017). Freezing foods. Colorado State University Extension. Retrieved from https://extension.colostate.edu/topic-areas/nutrition-food-safety-health/freezing-foods-9-346/
5. Chatterjee, R. N., & Rajkumar, U. (2015). An overview of poultry production in India. *Indian Journal of Animal Health*, *54*(2), 89-108.
6. Delgado, D., & Hanafy, A. (2020). Importance of Freeze-Drying in Pharmaceuticals, Food, and Biotechnology Industries. AIMS Bioengineering, 7(3), 118-138.
7. Fellows, P. (2009). Food processing technology: Principles and practice (3rd ed.). Woodhead Publishing.
8. Fellows, P. (2009). Food processing technology: Principles and practice (3rd ed.). Boca Raton, FL: CRC Press.
9. Fellows, P. (2009). Food processing technology: Principles and practice (3rd ed.). Boca Raton, FL: CRC Press.
10. Fellows, P. (2009). Food processing technology: Principles and practice (3rd ed.). Woodhead Publishing.
11. Jay, J. M., Loessner, M. J., & Golden, D. A. (2008). *Modern food microbiology*. Springer Science & Business Media.
12. Jilo, S. A., & Hasan, L. A. The Importance of Poultry Meat in Medicine: A.
13. Kerry, J. P., & Kerry, J. F. (2002). Meat Processing: Improving Quality. CRC Press.

14. Lafarga, T., & Hayes, J. E. (2017). Liquid smoke flavorings: Current and future applications for improving food quality. Comprehensive Reviews in Food Science and Food Safety, 16(4), 472-488. doi: 10.1111/1541-4337.12263

15. Lafarga, T., & Hayes, J. E. (2017). Liquid smoke flavorings: Current and future applications for improving food quality. Comprehensive Reviews in Food Science and Food Safety, 16(4), 472-488. doi: 10.1111/1541-4337.12263

16. Lafarga, T., & Hayes, J. E. (2017). Smoke flavorings in food: A review. Journal of Food Science, 82(6), 1575-1584. https://doi.org/10.1111/1750-3841.13730

17. Laguerie, C., & Chen, X. D. (2018). Vacuum Drying. In Laguerie, C., Chen, X. D., Bhandari, B., & Delmas, H. (Eds.), Drying Technologies in Food Processing (pp. 55-74). John Wiley & Sons.

18. Mahmoud, D. H., Mahmoud, H. B., EL-Nawawi, F. A. M., & Abdel-Naeem, H. H. S. (2021). Impact of thawing methods on the bacteriological quality of chicken meat. International Journal of Veterinary Science, 10(3), 214-219.

19. Mead, G. (Ed.). (2004). *Poultry meat processing and quality*. Woodhead Publishing.

20. Mertens, B., & Luyten, M. (2019). The Process of Freeze-Drying: A Review. CryoLetters, 40(6), 321-334.

21. Nakajima, M., Dokko, Y., Mori, M., & Sakamoto, J. (2019). Psychological effects of smoking on the mind and body. Tobacco Induced Diseases, 17(September). https://doi.org/10.18332/tid/110188

22. National Center for Home Food Preservation. (n.d.). Drying Fruits and Vegetables

23. National Center for Home Food Preservation. (n.d.). Principles of Home Canning. University of Georgia, College of Family and Consumer Sciences.

24. OECD (2018) Market examinations in Mexico: Case study of the chicken meat market.

25. Pal, M., Ayele, Y., Hadush, A., Kundu, P., & Jadhav, V. J. (2017). Journal of Experimental Food Chemistry.

26. Rhee, M. S., Lee, S. Y., Lee, H. G., & Oh, D. H. (2013). Effectiveness of refrigeration in inhibiting growth of Escherichia coli O157:H7,

Salmonella spp., and Staphylococcus aureus in ground beef. Journal of Food Protection, 76(12), 2114-2118.

27. Ruiz-Carrascal, J. (2010). Curing of meat products. In Handbook of Meat Processing (pp. 227-241). Blackwell Publishing Ltd.
28. Ruiz-Carrascal, J. (2010). Curing. In J. W. Park & F. Toldrá (Eds.), Handbook of meat processing (pp. 111-130). Blackwell Publishing Ltd.
29. Sánchez-Escalante, A., Torrescano, G., & Mota-Rojas, D. (2011). A review of natural and synthetic smoke generation for the food industry. In S. A. Khan & A. S. Malik (Eds.), Research advancements in food and nutrition sciences (pp. 47-57). New York, NY: Nova Science Publishers.
30. Sánchez-Escalante, A., Torrescano, G., & Mota-Rojas, D. (2011). A review of natural and synthetic smoke generation for the food industry. In S. A. Khan & A. S. Malik (Eds.), Research advancements in food and nutrition sciences (pp. 47-57). New York, NY: Nova Science Publishers.
31. Sánchez-Escalante, A., Torrescano, G., & Mota-Rojas, D. (2011). Smoking as a method for food preservation: A review. International Journal of Food Science and Technology, 46(11), 2252-2265.
32. Scanes, C. G. (2007). The global importance of poultry. *Poultry science*, *86*(6), 1057.
33. Schirmer, B. C., & Lorenzen, C. L. (2019). Rapid cooling of beef and pork carcasses using different blast and air chilling methods: Impact on meat quality and shelf life. Meat Science, 148, 48-57.
34. Sharma, H. P., Mendonca, A. F., & Demirci, A. (2016). Curing of meat products: Review. Journal of Food Science and Technology, 53(9), 4064-4079.
35. Singh, A. K. (2014). Freezing as a food preservation method. Food Science and Technology, 4(2), 22-27. doi: 10.12691/ijfst-2-4-1
36. United States Department of Agriculture. (2017). Meat and Poultry Labeling Terms.
37. United States Department of Agriculture. (2020). Complete Guide to Home Canning.
38. United States Department of Agriculture. (n.d.). Meat & Poultry Hazards and Controls Guide.
39. Wahyono, N. D., & Utami, M. M. D. (2018). A review of the poultry meat production industry for food safety in Indonesia. In *Journal of Physics: conference series* (Vol. 953, No. 1, p. 012125). IOP Publishing.

Chapter - 15

Methods of Cooking; Frying, Roasting, Broiling, Braising and Stewing

<u>Authors</u>

Mamta

Ph.D. Scholar, Department of Food Science, Nutrition and Technology, CSK Himachal Pradesh Agriculture University, Palampur, Himachal Pradesh, India

Vaishnavi

Ph.D. Scholar, Department of Food Science, Nutrition and Technology, CSK Himachal Pradesh Agriculture University, Palampur, Himachal Pradesh, India

Pooja Bhatt

Assistant Professor, Chandigarh College of Hospitality, CGC, Landran, Punjab, India

Nivedhika Rai Vaid

Assistant Professor, Chandigarh College of Hospitality, CGC, Landran, Punjab, India

Chapter - 15

Methods of Cooking; Frying, Roasting, Broiling, Braising and Stewing

Mamta, Vaishnavi, Pooja Bhatt and Nivedhika Rai Vaid

Abstract

People have been cooking since the very beginning of time. Roti, puris, paranthas, rice, pulao, lentils, veggies, salad, chutney, pickles, and curd are just a few of the numerous foods we consume on a regular basis. Because they were all prepared differently, all foods taste differently from one another. In order to meet the family's nutritional demands, food preparation is a vital step. Food must be aesthetically pleasing and palatable in order to be ingested. Nuts, fruits, and vegetables can all be eaten raw, but for most meals to be effective, they must be prepared. Food is cooked when it is subjected to heat's effects. The nutritional and physio-chemical properties of macro- and micronutrients are also impacted by cooking. Radiation, convection, and conduction all contribute to the transfer of heat to the food during cooking. The type of cooking technique used relies on how heat is applied to the food during cooking; there are several different techniques, including moist heat, dry heat, combination, and others.

15.1 Definition, aims and objectives of cooking food

The preparation of food is a crucial stage in providing for the family's nutritional needs. In order to be consumed, food must be aesthetically and gastronomically appealing. Nuts, fruits, and vegetables can all be consumed raw, but most meals must be prepared in order to achieve the desired results. Cooking is the process of exposing food to the effects of heat.

1) Food is partially sterilized by cooking. Bacterial growth slows significantly and eventually stops above 45°C (113°F) above 40°C (104°F). For example, milk is pasteurized at 63°C (145°F) for 30 minutes or at 72°C (161°F) for 15 seconds to make it safe. Non-sporing bacteria are killed at temperatures above 60°C (140°F) for varied durations of time. With the exception of spores, boiling instantly kills living cells. It takes 4 to 5 hours of boiling to kill germs

that have spores. Higher heat must be used in order to destroy them faster. Cooking helps to make food more digestible. Complex foods are often split into simpler substances during cooking. This helps the body to absorb and utilize the food more readily than if consumed in its raw form.

2) Cooking makes food more palatable.

3) Food that has been cooked looks better and is consequently more appetizing.

4) Cooking offers diversity. The same component can be used to produce a wide variety of recipes.

5) Food becomes easier to digest when it is cooked. During cooking, complicated meals are frequently broken down into simpler ingredients. When compared to eating the meal raw, this makes it easier for the body to absorb and use the food.

A balanced supper is easier to offer when you cook. Making a dish with a variety of ingredients makes it simpler to serve a balanced dinner.

15.2 Effects of heat on food- carbohydrates, proteins, fats, vitamins, minerals

1) Action of heat on carbohydrates

When cooked in an aqueous media, monosaccharide, oligosaccharides, and polysaccharides go through a variety of changes. Over 100 chemicals are created as a result of the breakdown and epimerization of the sugars. These substances don't have any negative consequences. When sugar is cooked in water, syrups of varying strengths can be created. Sugar cookery can also reach the varied balls stage (soft, hard, and brittle). When heated in an aqueous or moist environment, starch molecules which are the principal source of calories in many diets swell, rupture, and burst. This causes starch to become gelatinized, which allows for more enzymatic digestion by enzymes like Amylases. Thus, cooking makes carbohydrates more palatable for digestion. When starch is exposed to dry heat at a temperature of 200°C or higher, it breaks down resulting in the formation of brown colored intermediate compound called dextrin and the process is called dextrinization. For example, toasting of bread.

2) Action of heat on protein

When heated, protein coagulates or sets. The egg white is a nice example. The coagulation process aids in retaining the food's nutrition inside, where it is sealed in by the development of an external covering. Consequently, when cooking items high in protein, like meat and fish, use hot water. Put the meat in an oven that is preheated to a moderate temperature before turning the heat down to roast it. The softening of muscle fibers is another effect of heat on meat. The elastin in connective tissue shrinks during normal cooking, and heat and water induce collagen to turn into gelatin, which causes the muscle fibers to separate and the meat to become soft. At a medium temperature compared to a high temperature, this is accomplished more successfully. As it cooks, meat shrinks as well. The shrinkage increases with increasing temperature. Foods containing protein become more digestible when cooked at medium temperatures. Meat is red because of the haemoglobin in the capillaries and myoglobin in the muscle tissues. Both of them break down when heated, giving cooked meat its characteristic brown colour. 65 °C (149 °F) is the temperature at which the transformation occurs. Since the protein itself is denatured at high temperatures, it loses some of its nutritional value

3) Action of heat on fats

When fat is exposed to heat, it melts. Fat break down into fatty acids and acrolein at high temperatures. High heat cooking, as well as the presence of carbonized breadcrumbs and other minute pieces of cooked food, all contribute to the darkening of fat. The smoke point of a fat is the temperature at which breakdown begins when the fat is heated. Repeated heating lowers the precise temperature at which this occurs, which varies depending on the type of fat. As a result of high heat, moisture, air, and the presence of carbonized crumbs and small food bits, fat degrades over time to the point that it will only bubble in the pan and is insufficient to carry out the task of frying. The breakdown is further aided by abnormal absorption brought on by cooking food at a temperature that is too low for too long. Additionally, fat starts to solidify and get gooey. Fat that has gone through this process, known as polymerization, is no longer suitable for usage.

4) Action of heat on minerals

The only significant mineral loss associated with cooking occurs when cooking liquid is discarded. Cooking makes some minerals readily available.

5) Action of heat on vitamins

The only vitamin that loses stability when heated is vitamin C, though some of it can be retained with careful cooking. Normal cooking techniques do not degrade vitamin A and D. When cooking at high temperatures, such as when making morning cereals or baking biscuits, adding baking soda or discarding cooking liquid, vitamin B may be damaged.

19.3 Methods of heat transfer- conduction, convection, radiation

Heat transfer

Heat must be delivered from the heat source (such as a gas flame or heating element coil) to and through the food in order for it to cook. The regulation of the cooking process is aided by an understanding of the manner and rate of heat transfer. There are three ways to transfer heat: conduction, convection, and radiation.

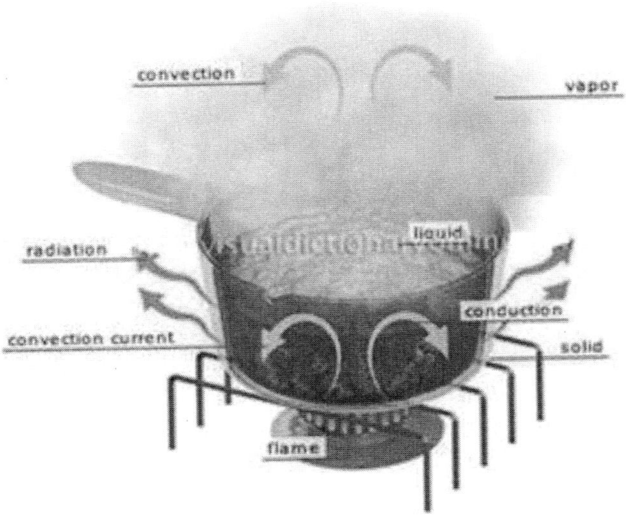

Conduction can happen in one of two ways:

1) When heat transfers immediately from an object to anything it touches.
2) When heat transfers from one area of an object to an area that is nearby.

The rate at which different materials transmit heat varies. Copper and aluminium conduct heat quickly, stainless steel more slowly, and glass and porcelain yet more slowly. Heat does not transmit well through air.

Convection: Convection is the process by which heat is dispersed through the movement of air, steam, or liquid (including hot fat). There are two types of convection:

1) **Natural:** Cooler gases and liquids sink while hotter ones rise. Thus, there is a continuous natural circulation that distributes the heat in any oven, kettle of water, or deep fat fryer.

2) **Mechanical:** Fans hasten the circulation of heat in steamers and convection ovens. As a result, the heat is distributed more quickly and evenly, which speeds up the cooking process.

Convection is mechanically effected by stirring. Natural circulation moves more slowly in thick liquids because they cannot move as quickly as thin ones.

Radiation: When energy is transported from the source to the food through waves, it is called radiation. Although the waves don't actually contain heat energy, they do when they hit the food being cooked.

In the kitchen, radiation is employed in two different ways:

1) **Infrared:** The most well-known application of infrared cooking is broiling. In a grill, a ceramic or electric element heated by a gas flame heats to such a high temperature that it emits infrared radiation, which cooks the food. High intensity infrared ovens are another option for quickly heating meals.

2) **Microwave:** When using a microwave oven, the radiation it produces partially penetrates the meal, agitating the water molecules there. The food is cooked by the immense heat produced by the friction this movement causes. A substance that is totally free of water will not warm up in the microwave because microwave radiation only impacts water molecules. Only the transfer of heat from the food causes plates to heat up. Additionally, since microwaves can only reach a depth of 2" into food, heat is conducted into the centre of huge portions.

19.4 Cooking methods-moist & dry

Cooking methods

During cooking, heat is transmitted to the food by conduction, convection, and radiation. The kind of cooking method utilised depends on how heat is given to the food during cooking. The methods created can be divided into four categories:

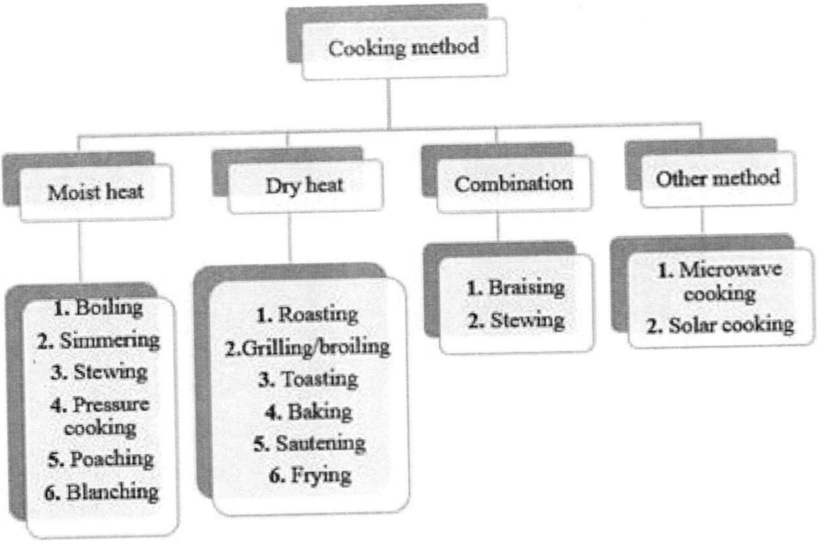

Classification of cooking methods

Moist heat methods

1) Boiling

It is a technique for preparing food that involves just submerging it in water that is 100°C and keeping it there until the food gets soft. For instance, boiling can be used to cook rice, eggs, dhal, meat, roots, and tubers.

2) Simmering

At a temperature just below boiling, food is cooked in liquid. When the liquid hits 180–190°C, little bubbles begin to appear at the bottom of the pot and occasionally break the surface. It is essential for cooking meals in a lot of water, such as pasta or soup.

3) Stewing

It refers to simmering food with little liquid, just enough to cover half of the meal, in a skillet with a tight-fitting lid. This style of cooking is slow. Once the liquid reaches boiling, the heat is turned down to simmering (82°C to 90°C) levels. The steam produced in the pan cooks the food that is above the liquid. Typical stewing ingredients include apple, meat, roots, veggies, and legumes.

4) Steaming

It is a technique for cooking food under steam produced by water that is vigorously boiling in a pan. The food that will be steam-cooked is put in a container so that it doesn't come into touch with any liquid or water. Steaming is used to prepare idiappam, custard, and idli. Veggies can also be steam-cooked.

5) **Pressure cooking**

Pressure cooking is a technique that uses steam under pressure, and a pressure cooker is the tool that is utilised. With this technique, boiling water can be heated above 100°C. For instance, you can pressure cook rice, dhal, meat, roots, and tubers.

6) **Poaching**

This involves cooking in a little amount of liquid at temperatures between 80°C–85°C, which are below the boiling point. Poached fish and eggs are two examples.

7) **Blanching**

When preparing a meal, it is frequently sufficient to just peel the skin off of fruits and vegetables without further preparation. Blanching can be used to accomplish this. According to the food's texture, food is submerged in boiling water for 5 seconds to 2 minutes. This assists in skin or peel removal without causing food to become softer. Blanching can also be accomplished by briefly submerging food in boiling water or by briefly exposing food to boiling heat before dropping it into cold water. The procedure results in loose skin that is simple to pull off.

Dry heat methods

1) Roasting

With this technique, food is roasted in an unprotected frying pan or heated griddle. However, roasting can be done with or without a cooking medium. For instance, you can roast groundnuts with or without oil.

2) Grilling/Broiling

Food is cooked by being exposed to direct heat and is referred to as grilling or broiling. Food is positioned beneath, above, or in between a surface that is extremely hot in this approach. For instance: fish, poultry, corn and chapatti.

3) **Toasting**

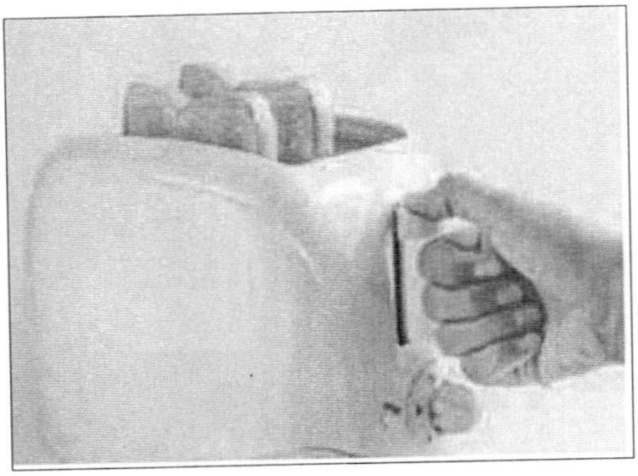

This technique allows for easy browning on all sides because the meal is kept between two heating devices. Toasted bread slices are an example.

4) **Baking**

In this technique, the meal is cooked using dry heat in an oven or gadget that resembles an oven. A typical oven maintains temperatures between 120°C and 260°C. The meal is typically kept exposed in a container lined with paper that has been greased with lard. This technique is used to make meat, bread, cake, biscuits, and pastries.

5) **Sauteing**

A technique called sautéing involves tossing food in just enough oil to coat the bottom of the pan. The flame or degree of heat is lowered and the pan is covered with a lid. The food is allowed to cook in its own steam until it is soft. To ensure that all the pieces contact the oil and are cooked evenly, the dish is occasionally tossed or rotated with a spatula. This approach yields a product that is soft and slightly moist but devoid of any liquid or gravy. Typically, sauteed foods are veggies that are served as side dishes on a menu. Sauteing can be used in combination with other techniques to create a variety of dishes.

6) Frying

With this technique, a greater quantity of heated fat is in contact with the food to be cooked. Deep fat frying is the term used when food is completely submerged in hot oil. Deep-fat fried food varieties include samosas, chips, and pakodas. Only a small amount of fat is used in shallow fat frying, and the meal is turned over to allow both sides to brown. Like omelettes, cutlets, and paranthas.

Combination of cooking method

1) Braising

Combining roasting and stewing in a skillet with a tight-fitting lid is known as braising. Food is given seasonings and flavors before being given time to cook gradually. Combination method food preparations include:

- Uppuma - Roasting and boiling.
- Cutlet - Boiling and deep frying.
- Vermicelli payasam - Roasting and simmering.

2) Stewing

This technique combines simmering and frying. For instance, a stew of meat.

Other methods

1) Microwave cooking

A magnetron tube is the source of electro-magnetic radiation with a high frequency wave, which cooks food in a microwave oven. Food should be stored in non-metallic, plastic, glass, or chinaware containers. Because they transmit microwaves rather than absorbing or reflecting them, these containers are employed. For instance, a cake can be baked in a microwave.

2) Solar cooking

Solar cooking is a fairly straightforward method that uses the sun's energy. A well-insulated box that is internally painted black and covered with one or more transparent coverings makes up a solar cooker. These covers permit solar radiation to enter the box but prevent heat from the heated, black absorption plate from leaving the box. Because of this, temperatures that are suitable for cooking can be reached up to 140°C. Solar cooking eliminates burning and content oozing as in rice.

References

1. https://uou.ac.in/sites/default/files/slm/BHM-201T.pdf
2. https://www.textbooksonline.tn.nic.in/Books/11/Nutrition-EM/CHAPTER_2.pdf